Beautiful Geometry

Frontispiece: *Infinity*

Beautiful Geometry

ELI MAOR and EUGEN JOST

Princeton University Press
Princeton and Oxford

Published by Princeton University Press, 41 William Street, Princeton, New Jersey 08540

In the United Kingdom: Princeton University Press, 6 Oxford Street, Woodstock, Oxfordshire OX20 1TW

press.princeton.edu

Jacket Illustration: *Sierpinski's Arrowhead* by Eugen Jost

Library of Congress Cataloging-in-Publication Data

Maor, Eli.

Beautiful geometry / Eli Maor and Eugen Jost.

pages cm

Summary: "If you've ever thought that mathematics and art don't mix, this stunning visual history of geometry will change your mind. As much a work of art as a book about mathematics, Beautiful Geometry presents more than sixty exquisite color plates illustrating a wide range of geometric patterns and theorems, accompanied by brief accounts of the fascinating history and people behind each. With artwork by Swiss artist Eugen Jost and text by acclaimed math historian Eli Maor, this unique celebration of geometry covers numerous subjects, from straightedge-and-compass constructions to intriguing configurations involving infinity. The result is a delightful and informative illustrated tour through the 2,500-year-old history of one of the most important and beautiful branches of mathematics"— Provided by publisher.

Includes index.

ISBN-13: 978-0-691-15099-4 (cloth : acid-free paper)

ISBN-10: 0-691-15099-0 (cloth : acid-free paper)   1. Geometry—History.   2. Geometry—History—Pictorial works.   3. Geometry in art.   I. Jost, Eugen, 1950–   II. Title.

QA447.M37 2014

516—dc23

2013033506

British Library Cataloging-in-Publication Data is available

This book has been composed in Baskerville 10 Pro

Printed in Canada

1 3 5 7 9 10 8 6 4 2

To Dalia, my dear wife of fifty years

May you enjoy many more years of
good health, happiness, and Naches from your family.

—Eli

To my dear Kathrin and to my whole family

Two are better than one; because they have a good reward for their labor.
For if they fall, the one will lift up his fellow (Ecclesiastes 4:9-10).

—Eugen

Contents

ART THROUGH MATHEMATICAL EYES

ELI MAOR

No doubt many people would agree that art and mathematics don't mix. How could they? Art, after all, is supposed to express feelings, emotions, and impressions—a subjective image of the world as the artist sees it. Mathematics is the exact opposite—cold, rational, and emotionless. Yet this perception can be wrong. In the Renaissance, mathematics and art not only were practiced together, they were regarded as complementary aspects of the human mind. Indeed, the great masters of the Renaissance, among them Leonardo da Vinci, Michelangelo, and Albrecht Dürer, considered themselves as architects, engineers, and mathematicians as much as artists.

If I had to name just one trait shared by mathematics and art, I would choose their common search for pattern, for recurrence and order. A mathematician sees the expression $a^2 + b^2$ and immediately thinks of the Pythagorean theorem, with its image of a right triangle surrounded by squares built on the three sides. Yet this expression is not confined to geometry alone; it appears in nearly every branch of mathematics, from number theory and algebra to calculus and analysis; it becomes a pattern, a paradigm. Similarly, when an artist looks at a wallpaper design, the recurrence of a basic motif, seemingly repeating itself to infinity, becomes etched in his or her mind as a pattern. *The search for pattern* is indeed the common thread that ties mathematics to art.

• ◆ •

The present book has its origin in May 2009, when my good friend Reny Montandon arranged for me to give a talk to the upper mathematics class of the Alte Kantonsschule (Old Cantonal High School) of Aarau, Switzerland. This school has a historic claim to fame: it was here that a 16-year-old Albert Einstein spent two of his happiest years, enrolling there at his own initiative to escape the authoritarian educational system he so much loathed at home. The school still occupies the same building that Einstein knew, although a modern wing has been added next to it. My wife and I were received with great honors, and at lunchtime I was fortunate to meet Eugen Jost.

I had already been acquainted with Eugen's exquisite mathematical artwork through our mutual friend Reny, but to meet him in person gave me special pleasure, and we instantly bonded. Our encounter was the spark that led us to collaborate on the present book. To our deep regret, Reny Montandon passed away shortly before the completion of our book; just one day before his death, Eugen spoke to him over the phone and told him about the progress we were making, which greatly pleased him. Sadly he will not be able to see it come to fruition.

Our book is meant to be enjoyed, pure and simple. Each topic—a theorem, a sequence of numbers, or an intriguing geometric pattern—is explained in words and accompanied by one or more color plates of Eugen's artwork. Most topics are taken from geometry; a few deal with numbers and numerical progressions. The chapters are largely independent of one another, so the reader can choose what he or she likes without affecting the continuity of reading. As a rule we followed a chronological order, but occasionally we grouped together subjects that are related to one another mathematically. I tried to keep the technical details to a minimum, deferring some proofs to the appendix and referring others to external sources (when referring to books already listed in the bibliography, only the author's name and the book's title are given). Thus the book can serve as an informal—and most certainly not complete—survey of the history of geometry.

Our aim is to reach a broad audience of high school and college students, mathematics and science teachers, university instructors, and laypersons who are not afraid of an occasional formula or equation. With this in mind, we limited the level of mathematics to elementary algebra and geometry ("elementary" in the sense that no calculus is used). We hope that our book will inspire the reader to appreciate the beauty and aesthetic appeal of mathematics and of geometry in particular.

Many people helped us in making this book a reality, but special thanks go to Vickie Kearn, my trusted editor at Princeton University Press, whose continuous enthusiasm and support has encouraged us throughout the project; to the editorial and technical staff at Princeton University Press for their efforts to ensure that the book meets the highest aesthetic and artistic standards; to my son Dror for his technical help in typing the script of plate 26 in Hebrew; and, last but not least, to my dear wife Dalia for her steady encouragement, constructive critique, and meticulous proofreading of the manuscript.

PLAYING WITH PATTERNS, NUMBERS, AND FORMS

EUGEN JOST

My artistic life revolves around patterns, numbers, and forms. I love to play with them, interpret them, and metamorphose them in endless variations. My motto is the Pythagorean motto: *Alles ist Zahl* ("All is Number"); it was the title of an earlier project I worked on with my friends Peter Baptist and Carsten Miller in 2008. *Beautiful Geometry* draws on some of the ideas expressed in that earlier work, but its conception is somewhat different. We attempt here to depict a wide selection of geometric theorems in an artistic way while remaining faithful to their mathematical message.

While working on the present book, my mind was often with Euclid: A point is that which has no part; a line is a breadthless length. Notwithstanding that claim, Archimedes drew his broad-lined circles with his finger in the sand of Syracuse. Nowadays it is much easier to meet Euclid's demands: with a few clicks of the mouse you can reduce the width of a line to nothing—in the end there remains only a nonexisting path. It was somewhat awe inspiring to go through the constructions that were invented—or should I say discovered?—by the Greeks more than two thousand years ago.

For me, playing with numbers and patterns always has top priority. That's why I like to call my pictures playgrounds, following a statement by the Swiss Artist Max Bill: "perhaps the goal of concrete art is to develop objects for mental use, just like people created objects for material use." Some illustrations in our book can be looked at in this sense. The onlooker is invited to play: to find out which rules a picture is built on and how the many metamorphoses work, to invent his or her own pictures. In some chapters the relation between text and picture is loose; in others, however, artistic claim stood behind the goal to enlighten Eli's text. Most illustrations were created on the keypad of my computer, but others are acrylics on canvas. Working with Eli was a lot of fun. He is one of those mathematicians that teach you: Mathematics did not fall from heaven; it was invented and found by humans; it is full of stories; it is philosophy, history and culture. I hope the reader will agree.

Beautiful Geometry

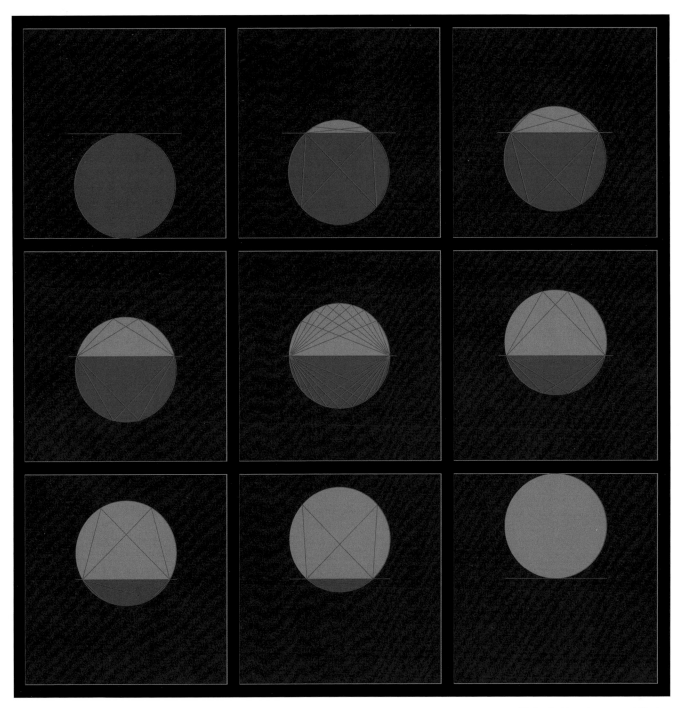

Plate 1. *Sunrise over Miletus*

1

Thales of Miletus

Thales (ca. 624–546 BCE) was the first of the long line of mathematicians of ancient Greece that would continue for nearly a thousand years. As with most of the early Greek sages, we know very little about his life; what we do know was written several centuries after he died, making it difficult to distinguish fact from fiction. He was born in the town of Miletus, on the west coast of Asia Minor (modern Turkey). At a young age he toured the countries of the Eastern Mediterranean, spending several years in Egypt and absorbing all that their priests could teach him.

While in Egypt, Thales calculated the height of the Great Cheops pyramid, a feat that must have left a deep impression on the locals. He did this by planting a staff into the ground and comparing the length of its shadow to that cast by the pyramid. Thales knew that the pyramid, the staff, and their shadows form two similar right triangles. Let us denote by H and h the heights of the pyramid and the staff, respectively, and by S and s the lengths of their shadows (see figure 1.1). We then have the simple equation $H/S = h/s$, allowing Thales to find the value of H from the known values of S, s, and h. This feat so impressed Thales's fellow citizens back home that

they recognized him as one of the Seven Wise Men of Greece.

Mathematics was already quite advanced during Thales's time, but it was entirely a practical science, aimed at devising formulas for solving a host of financial, commercial, and engineering problems. Thales was the first to ask not only *how* a particular problem can be solved, but *why*. Not willing to accept facts at face value, he attempted to prove them from fundamental principles. For example, he is credited with demonstrating that the two base angles of an isosceles triangle are equal, as are the two vertical angles formed by a pair of intersecting lines. He also showed that the diameter of a circle cuts it into two equal parts, perhaps by folding over the two halves and observing that they exactly overlapped. His proofs may not stand up to modern standards, but they were a first step toward the kind of deductive mathematics in which the Greeks would excel.

Thales's most famous discovery, still named after him, says that from any point on the circumference of a circle, the diameter always subtends a right angle. This was perhaps the first known *invariance* theorem—the fact that in a geometric configuration,

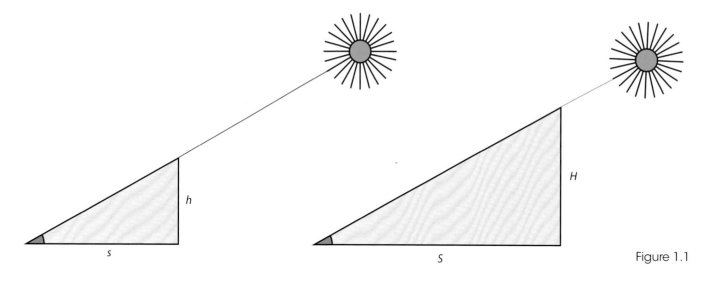

Figure 1.1

some quantities remain the same even as others are changing. Many more invariance theorems would be discovered in the centuries after Thales; we will meet some of them in the following chapters.

Thales's theorem can actually be generalized to any chord, not just the diameter. Such a chord divides the circle into two unequal arcs. Any point lying on the larger of these arcs subtends the chord at a constant angle $\alpha < 90°$; any point on the smaller arc subtends it at an angle $\beta = 180° - \alpha > 90°$. [1] Plate 1, *Sunrise over Miletus*, shows this in vivid color.

NOTE:

1. The converse of Thales's theorem is also true: the locus of all points from which a given line segment subtends a constant angle is an arc of a circle having the line segment as chord. In particular, if the angle is 90°, the locus is a full circle with the chord as diameter.

2

Triangles of Equal Area

Around 300 BCE, Euclid of Alexandria wrote his *Elements*, a compilation of the state of mathematics as it was known at his time. Written in 13 parts ("books") and arranged in strict logical order of definitions, postulates (today we call them axioms), and propositions (theorems), it established mathematics as a *deductive* discipline, in which every theorem must be proved based on previously established theorems, until we fall back on a small set of postulates whose validity we assume to be true from the outset. Euclid's 23 opening definitions, 10 axioms, and 465 theorems cover all of classical ("Euclidean") geometry—the geometry we learn in school—as well as elementary number theory. The *Elements* is considered the most influential book in the history of mathematics. It has had an enormous influence on generations of mathematicians, scientists, and philosophers, and its terse style and rigid structure of definitions, postulates, propositions, and demonstrations (proofs) became the paradigm of how mathematics should be done.

Proposition 38 of Book I of the *Elements* says, *Triangles which are on equal bases and in the same parallels are equal to one another*. In modern language this reads: all triangles with the same base and top vertices that lie on a line parallel to the base have the same area.

In proving this theorem, it would be tempting to use the familiar formula for the area of a triangle, $A = bh/2$, and argue that since two triangles with the same base b and top vertices that lie on a line parallel to the base also have the same height h, they must have the same area. This "algebraic" proof, however, would not satisfy Euclid; he insisted that a proof should be based strictly on geometric considerations. Here is how he proved it:

In figure 2.1, let triangles *ABC* and *DEF* have equal bases *BC* and *EF*. Their vertices *A* and *D* are on a line parallel to *BC* and *EF*. We extend *AD* to points *G* and *H*, where *GB* is parallel to *AC* and *HF* is parallel to *DE*. Then the figures *GBCA* and *DEFH* are parallelograms with the same area, for they have equal bases *BC* and *EF* and lie between the same parallels *BF* and *GH*. Now the area of triangle *ABC* is half the area of parallelogram *GBCA*, and the area of triangle *DEF* is half the area of parallelogram *DEFH*. Therefore the two triangles have the same area— QED.[1]

4

Plate 2. *Triangles of Equal Area*

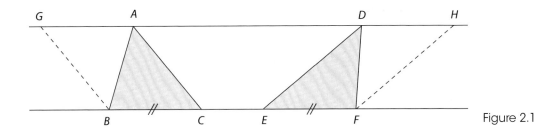

Figure 2.1

Our illustration (plate 2) shows three identical red triangles, each of whose sides can be regarded as a base. A series of blue triangles are built on each base, with their vertices moving along a line parallel to that base. They get narrower as the vertices move farther out, yet they all have exactly the same area, providing another example of a quantity that remains unchanged even as other quantities in the configuration vary.

This theorem may seem rather unassuming, but Euclid makes good use of it in proving other, more advanced theorems; most famous among them is the Pythagorean theorem, to which we will turn in chapters 5 and 6.

NOTE:

1. QED stands for *quod erat demonstrandum*, Latin for "that which was to be demonstrated."

3

Quadrilaterals

Here is a little-known jewel of a theorem that never fails to amaze me: take any quadrilateral (four-sided polygon), connect the midpoints of adjacent sides, and—surprise—you'll get a parallelogram! The surprise lies in the word *any*. No matter how skewed your quadrilateral is, the outcome will always be a parallelogram. The theorem holds true even for the dart-shaped quadrilateral shown in the top-left corner of plate 3. And that's not all: the area of the parallelogram will always be one half the area of the quadrilateral from which it was generated.

The proof is rather short and is based on the following theorem: in any triangle, the line joining the midpoints of two sides is parallel to the third side and is half as long (see figure 3.1). Now let's apply this to quadrilateral *ABCD*. Denote the midpoints of sides *AB*, *BC*, *CD* and *DA* by *P*, *Q*, *R*, and *S*, respectively (figure 3.2). Line *PQ* is parallel to diagonal *AC*, which in turn is parallel to *RS*. Thus, *PQ* and *RS* are parallel. By the same argument, lines *PS* and *RQ* are also parallel, so *PQRS* is a parallelogram—QED. (The proof that it has half the area of the generating quadrilateral is just a tad longer and is given in the appendix.)

Now, of course, you can repeat the process and connect the midpoints of *PQRS* to get another, smaller parallelogram, as shown in the middle panel of plate 3. In fact, you can do this again and again, getting ever smaller parallelograms whose areas are ½, ¼, ⅛, ... of the original quadrilateral, until they seem to converge to a point.

While we are on the subject of quadrilaterals, here is another little-known fact: the area of any quadrilateral is completely determined by the lengths of its two diagonals and the angle between them. In fact, the area is given by the simple formula $A = \frac{1}{2} d_1 \cdot d_2 \cdot \sin \alpha$ (figure 3.3). It doesn't matter how you

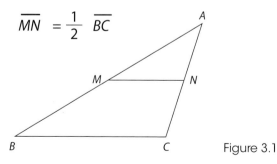

$$\overline{MN} = \frac{1}{2}\ \overline{BC}$$

Figure 3.1

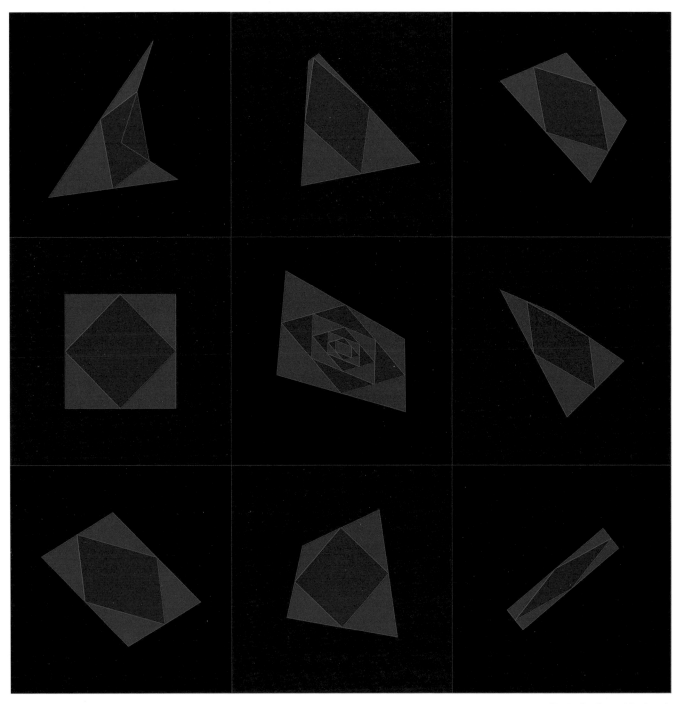

Plate 3. *Quadrilaterals*

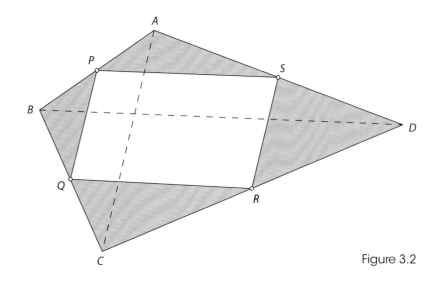

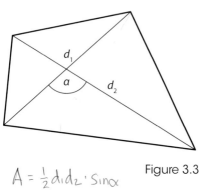

Figure 3.2

Figure 3.3

$$A = \frac{1}{2} d_1 d_2 \cdot \sin \alpha$$

measure the angle—α or $(180° - \alpha)$—for we know from trigonometry that $\sin \alpha = \sin (180° - \alpha)$. It is a pity that these little treasures seldom, if ever, find their way into our geometry textbooks.

Sin 45 = 0.707 = $\frac{\sqrt{2}}{2}$ * Sin of Supp #'s are =

Sin 135 = 0.707 = $\frac{\sqrt{2}}{2}$

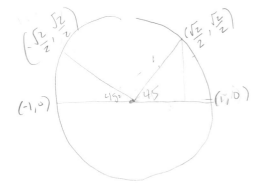

Cos 45 = 0.707 = $\frac{\sqrt{2}}{2}$ * Cos of Supp #'s are

Cos 135 = -0.707 = $\frac{-\sqrt{2}}{2}$ opposites

Sinα = Cos(90-α)   * The sin of an $\angle$ is
                                = the cos of the comp
                                     of the $\angle$

4

Perfect Numbers and Triangular Numbers

The Pythagoreans—the school founded by Pythagoras in the fifth century BCE—had a special relationship with numbers (the term here meaning positive integers). In their mind, numbers were not just a measure of quantity but symbols possessing mythical significance. The number 1 was not considered a number at all, but rather the generator of all numbers, since every number can be obtained from it by repeated addition. Two symbolized the female character, 3 the male character, and 5 their union. Five was also the number of Platonic solids—convex polyhedra whose faces are all identical regular polygons (although the proof that there are exactly 5 of them came only later). These 5 solids are the *tetrahedron*, having 4 equilateral triangles as faces, the *cube* (6 square faces), the *octahedron* (2 square pyramids joined at their bases and comprising 8 equilateral triangles), the *do-*decahedron (12 regular pentagons), and the *icosahedron* (20 equilateral triangles); they are shown in figure 4.1. No wonder, then, that the number 5 acquired something of a sacred status with the Greeks.

Even more revered than five was the number 6, the first *perfect number*, being the sum of its proper divisors, $1+2+3$.[1] The next perfect number is 28 ($=1+2+4+7+14$), followed by 496 and 8,128. These were the only perfect numbers known in antiquity. As of this writing, 48 perfect numbers are known; the largest, discovered in 2013, is $2^{57,885,160} \cdot (2^{57,885,161}-1)$, an enormous number of nearly 35 million digits. The question of how many perfect numbers exist— or even whether their number is finite or infinite—is still unanswered.

Six is also a *triangular number*, so called because these numbers form a triangular pattern when ar-

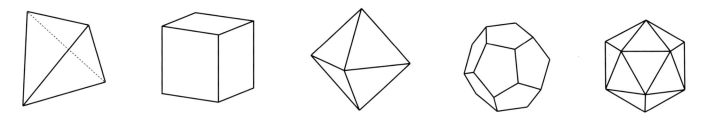

Figure 4.1

Plate 4. *Figurative Numbers*

ranged in rows of 1, 2, 3, ... dots. The first four triangular number are 1, $1+2=3$, $1+2+3=6$, and $1+2+3+4=10$, followed by 15, 21, and so on. The Pythagoreans discovered several relations between these sequences of integers. For example, the nth triangular number is always equal to $n(n+1)/2$ (you can check this for a few cases: $1+2+3=6=(3\times4)/2$, $1+2+3+4=10=(4\times5)/2$, etc.). So this gives us a convenient way—a formula—for finding the sum of the first n integers without actually adding them up:

$$1+2+3+\cdots+n=\frac{n(n+1)}{2}.$$

When we add instead the first n *odd* integers, a surprise is awaiting us: the result is always a perfect square: $1=1^2$, $1+3=4=2^2$, $1+3+5=9=3^2$, and, in general,

$$1+3+5+7+\cdots+(2n-1)=n^2.$$

Still another relation comes from adding two *consecutive triangular numbers*; again you get a perfect square: $1+3=4=2^2$, $6+10=16=4^2$, $10+15=25=5^2$, and so on. This is true because

$$\frac{n(n+1)}{2}+\frac{(n+1)(n+2)}{2}=(n+1)^2.$$

Perhaps most surprising of all is the fact that *every perfect number is also a triangular number*. Thus 6, 28, 496, and 8,128 are the 3rd, 7th, 31st, and 127th triangular numbers, respectively, and $2^{57,885,160}\cdot(2^{57,885,161}-1)$ is the $(2^{57,885,161}-1)$th triangular number.[2] Euclid, in his *Elements* (see page 3), proved that if 2^n-1 is prime, then $2^{n-1}\cdot(2^n-1)$ is perfect.[3] More than two thousand years later, Leonhard Euler proved the converse: every *even* perfect number is of the form $2^{n-1}\cdot(2^n-1)$ for some prime value of n. All 48 perfect numbers known today are even; whether any *odd*

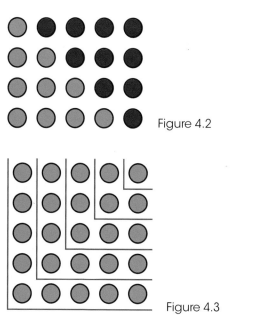

Figure 4.2

Figure 4.3

perfect numbers exist is unknown and remains one of the great mysteries of mathematics. Should such a number be found, it would be an oddity indeed!

The Pythagoreans established these relations, and many others, by representing numbers as dots arranged in various geometric patterns. For example, figure 4.2 shows two triangular arrays, each representing the sum $1+2+3+4$. Taken together, they form a rectangle of $4\times5=20$ dots. Therefore, the required sum is half of that, or 10. Repeating this for other numbers of dots, it would have been easy for the Pythagoreans to arrive at the formula $1+2+3+\cdots+n=n(n+1)/2$. Similarly, figure 4.3 illustrates how they would have established the formula $1+3+5+7+\cdots+(2n-1)=n^2$, while figure 4.4 demonstrates that the sum of two consecutive triangular numbers is always a perfect square. The Pythagoreans viewed these relations as purely geometric; today, of course, we prefer to prove them algebra-

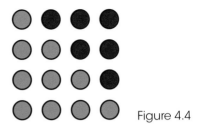

Figure 4.4

while others are artistic expressions of what a keen eye can discover in an assembly of dots. Note, in particular, the second panel in the top row: it illustrates the fact that the sum of eight identical triangular numbers, plus 1, is always a perfect square.[4]

ically. Yet, in discovering them, the Greeks sowed the seeds that many centuries later would evolve into modern number theory, the branch of mathematics concerned with the positive integers.

Plate 4, *Figurative Numbers*, is a playful meditation on ways of arranging 49 dots in different patterns of color and shape. Some of these arrangements hint at the number relations we mentioned previously,

NOTES:

1. The proper divisors of a number are all positive integers that divide it evenly, including 1 but excluding the number itself.

2. To see this, write $2^{57,885,160} \cdot (2^{57,885,161} - 1)$ as $[2^{57,885,161} \cdot (2^{57,885,161} - 1)]/2$ and let $n = 2^{57,885,161} - 1$. Then the expression has the form $(n+1)n/2$, a triangular number.

3. For more on primes of the form $2^n - 1$, see chapter 14.

4. This is because $8 \cdot (n+1)n/2 + 1 = 4n^2 + 4n + 1 = (2n+1)^2$.

5

The Pythagorean Theorem I

By any standard, the Pythagorean theorem is the most well-known theorem in all of mathematics. It shows up, directly or in disguise, in almost every branch of it, pure or applied. It is also a record breaker in terms of the number of proofs it has generated since Pythagoras allegedly proved it around 500 BCE. And it is the one theorem that almost everyone can remember from his or her high school geometry class.

Most of us remember the Pythagorean theorem by its famous equation, $a^2 + b^2 = c^2$. The Greeks, however, thought of it in purely geometric terms, as a relationship between areas. This is how Euclid stated it: *in all right-angled triangles the square on the side subtending the right angle is equal to the squares on the sides containing the right angle*. That is, the area of the square built on the hypotenuse ("the side subtending the right angle") is equal to the combined area of the squares built on the other two sides.

Pythagoras of Samos (ca. 580–ca. 500 BCE) may have been the first to prove the theorem that made his name immortal, but he was not the first to *discover* it: the Babylonians, and possibly the Chinese, knew it at least twelve hundred years before him, as is clear from several clay tablets discovered in Mesopotamia. Furthermore, if indeed he had a proof, it is

lost to us. The Pythagoreans did not leave any written records of their discoveries, so we can only speculate what demonstration he gave. There is, however, an old tradition that ascribes to him what became known as the *Chinese proof*, so called because it appeared in an ancient Chinese text dating from the Han dynasty (206 BCE – 220 CE; see figure 5.1). It is perhaps the simplest of the more than 400 proofs that have been given over the centuries.

The Chinese proof is by dissection. Inside square *ABCD* (figure 5.2) inscribe a smaller, tilted square *KLMN*, as shown in (a). This leaves four congruent right triangles (shaded in the figure). By reassembling these triangles as in (b), we see that the remaining (unshaded) area is the sum of the areas of squares 1 and 2, that is, the squares built on the sides of each of the right triangles.

Elisha Scott Loomis (1852–1940), a high school principal and mathematics teacher from Ohio, collected all the proofs known to him in a classic book, *The Pythagorean Proposition* (first published in 1927, with a second edition in 1940, the year of his death). In it you can find a proof attributed to Leonardo da Vinci, another by James A. Garfield, who would become the twentieth president of the United States, and yet another by Ann Condit, a

14

Plate 5. *25 + 25 = 49*

16-year-old high school student from South Bend, Indiana. And of course, there is the most famous proof of them all: Euclid's proof. We will look at it in the next chapter.

Our illustration (plate 5) shows a 45-45-90-degree triangle with squares—or what looks like squares—built on its sides and on the hypotenuse. But wait! Something strange seems to be going on: $5^2 + 5^2 = 7^2$, or $25 + 25 = 49$! Did anything go wrong? Do we see here an optical illusion? Not really: the illustration is, after all, an artistic rendition of the Pythagorean theorem, not the theorem itself; as such it is not bound by the laws of mathematics. To quote the American artist Josef Albers (1888–1976): "In science, one plus one is always two; in art it can also be three or more."

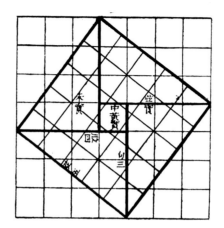

Figure 5.1. Joseph Needham, *Science and Civilisation in China*, courtesy of Cambridge University Press, Cambridge, UK.

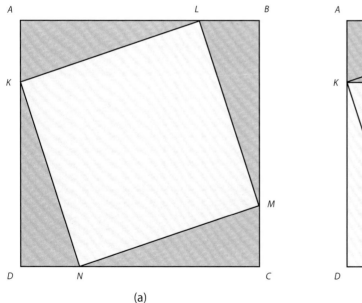

(a)

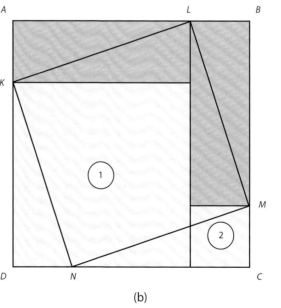

(b)

Figure 5.2

6

The Pythagorean Theorem II

The Pythagorean theorem is listed as Proposition 47 in the first book of Euclid's *Elements*. But you will not find Pythagoras's name heading it: true to his terse, matter-of-fact style, Euclid avoided any reference to persons in his work, instead letting the geometry speak for itself. So the most famous theorem in mathematics simply became known as Euclid I 47.

Euclid's proof of I 47 is anything but simple, and it has tested the patience of generations of students. In the words of philosopher Arthur Schopenhauer, "lines are drawn, we know not why, and it afterwards appears they were traps which close unexpectedly and take prisoner the assent of the astonished reader." Yet of the 400 or so demonstrations of the theorem, Euclid's proof stands out in its sheer austerity, relying on just a bare minimum of previously established theorems. Its classic configuration, with its many auxiliary lines, has become an icon in nearly every geometry book for the past one thousand years (see figure 6.1).

At the heart of Euclid's proof is a double application of theorem I 38 about triangles of equal area (see page 3). But first he proves a lemma (a preliminary result): the square built on one side of a right

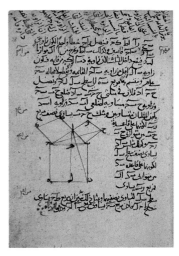

Figure 6.1. The Pythagorean Theorem in an Arab Text, from the Eighth Century. Richard Mankiewicz, *The Story of Mathematics*, courtesy of Princeton University Press, Princeton, NJ.

triangle has the same area as the rectangle formed by the hypotenuse and the projection of that side on the hypotenuse. Figure 6.2 shows a right triangle *ACB* with its right angle at *C*. Consider the square *ACHG* built on side *AC*. Project this side on the hypotenuse *AB*, giving you segment *AD*. Now construct

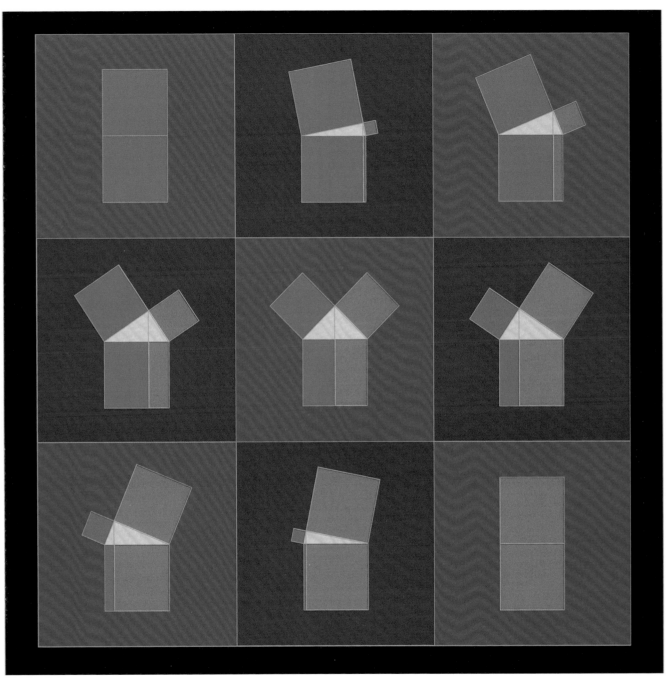

Plate 6. *Pythagorean Metamorphosis*

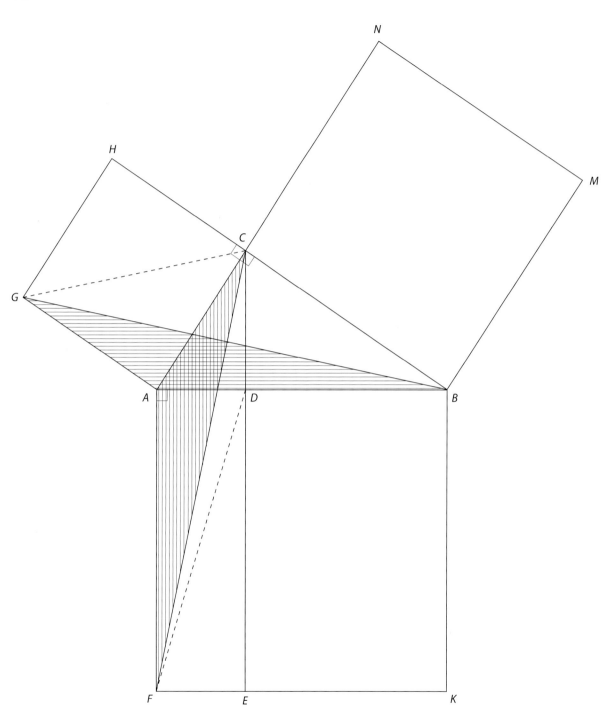

Figure 6.2

AF perpendicular to *AB* and equal to it in length. Euclid's lemma says that area *ACHG* = area *AFED*.

To show this, divide *AFED* into two halves by the diagonal *FD*. By I 38, area *FAD* = area *FAC*, the two triangles having a common base *AF* and vertices *D* and *C* that lie on a line parallel to *AF*. Likewise, divide *ACHG* into two halves by diagonal *GC*. Again by I 38, area *AGB* = area *AGC*, *AG* serving as a common base and vertices *B* and *C* lying on a line parallel to it. But area *FAD* = ½ area *AFED*, and area *AGC* = ½ area *ACHG*. Thus, if we could only show that area *FAC* = area *BAG*, we would be done.

It is here that Euclid produces his trump card: triangles *FAC* and *BAG* are congruent because they have two pairs of equal sides (*AF* = *AB* and *AG* = *AC*) and equal angles ∠*FAC* and ∠*BAG* (each con-sisting of a right angle and the common angle ∠*BAC*). And as congruent triangles, they have the same area.

Now, what is true for one side of the right triangle is also true of the other side (again, see figure 6.2): area *BMNC* = area *BDEK*. Thus, area *ACHG* + area *BMNC* = area *AFED* + area *BDEK* = area *AFKB*: the Pythagorean theorem.

Plate 6, *Pythagorean Metamorphosis*, shows a series of right triangles (in white) whose proportions change from one frame to the next, starting with the extreme case where one side has zero length and then going through several phases until the other side diminishes to zero. In accordance with Euclid's lemma, the two blue regions in each phase have equal areas, as do the orange regions.

7

Pythagorean Triples

A triple of positive integers (a, b, c) such that $a^2 + b^2 = c^2$ is called a *Pythagorean triple*; it represents a right triangle with sides a and b and hypotenuse c, all of integer lengths. Some examples are (3, 4, 5), (5, 12, 13), and (8, 15, 17); one can find such triples even among large numbers: (4,601, 4,800, 6,649). These four examples are of *primitive* triples— triples whose members have no common factor other than 1. Of course, from any given triple we can generate infinitely many others by multiplying it by an arbitrary integer; for example, the triple (6, 8, 10) is just the triple (3, 4, 5) magnified by a factor of 2. Such *nonprimitive* triples represent similar triangles and are essentially equivalent.

A Babylonian clay tablet known as Plimpton 322 and dating to about 1800 BCE (now at Columbia University) lists the hypotenuse and the short side of 15 Pythagorean triangles, demonstrating that the Babylonians were familiar with the Pythagorean theorem some twelve hundred years before Pythagoras is said to have proved it.[1] In Book X of the *Elements*, Euclid gives an algorithm for generating every primitive Pythagorean triple (there are infinitely many of them); we give it in the appendix.

It is hard to imagine a simpler geometric structure than a right triangle with integer sides, yet even this

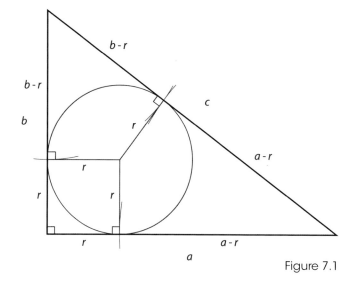

Figure 7.1

simple configuration holds within it some surprises. Take any right triangle and inscribe in it its *incircle*. This circle is tangent to all three sides, and its radius is given by the formula

$$r = \frac{a + b - c}{2}.$$

This follows from figure 7.1: because the two tangent lines from a point to a circle are of equal length,

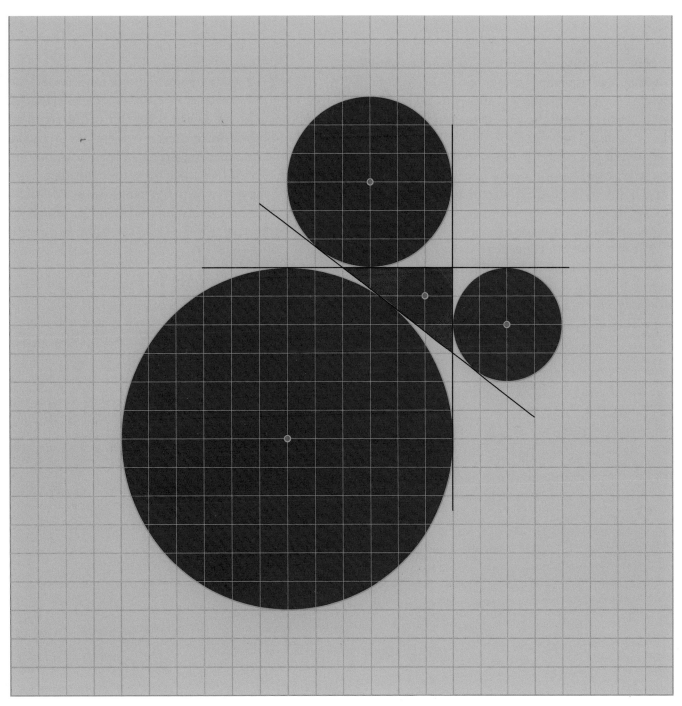

Plate 7. *The (3, 4, 5) Triangle and Its Four Circles*

we have $c = (a - r) + (b - r)$, from which the preceding formula follows.

Now, this formula works for any right triangle, whether Pythagorean or not; but if the triangle is Pythagorean, the radius will always be an integer as well. This is because in any Pythagorean triple, either a, b, and c are even, as in the triple $(6, 8, 10)$, or one of a or b is even, the other odd, and c is odd, as in $(9, 12, 15)$ [in a primitive triple, the latter case always holds, as in $(3, 4, 5)$]. In either case, $a + b - c$ will always be even and, therefore, divisible by 2, resulting in an integer value of r.

But that's not all. Consider the three *excircles* of a Pythagorean triangle, each being externally tangent to one side and to the other two sides *extended*. Surprisingly, the radii of all three circles are also integers. They are given by the formulas

$$r_a = \frac{c + a - b}{2}, \quad r_b = \frac{c + b - a}{2}, \quad r_c = \frac{c + b + a}{2}.$$

There exist some interesting relations among the four radii; for example, $r + r_a + r_b + r_c = a + b + c$, that is, their sum is equal to the perimeter of the triangle. Another peculiar relation is $r_a \cdot r_b = r \cdot r_c = (a \cdot b)/2$, the area of the triangle. The proofs of these relations are quite simple, and we leave them to the reader.

Plate 7 shows the $(3, 4, 5)$ triangle (in red) with its incircle and three excircles (in blue), for which $r = (3 + 4 - 5)/2 = 1$, $r_a = (5 + 3 - 4)/2 = 2$, $r_b = (5 + 4 - 3)/2 = 3$, and $r_c = (5 + 4 + 3)/2 = 6$.

NOTE:

1. See Maor, *The Pythagorean Theorem: A 4,000-Year History*, chapter 1.

8

The Square Root of 2

One of the most momentous events in the history of mathematics was the discovery of a new kind of number that had never been known before—an *irrational number*.

To the Pythagoreans, "number" meant either a positive integer or a ratio of two positive integers, a *rational number*. Examples of such numbers are ⅔ (or simply 2), ³⁄₂, and ⅗. The Pythagoreans believed that any quantity, whether an abstract number or a physical entity, is represented by a rational number. This belief, in all likelihood, came from music, a discipline that in ancient Greece ranked equal in importance to arithmetic, geometry, and spherics (astronomy)—the four components of the *quadrivium* that an educated person was expected to master. Pythagoras is said to have discovered that the common musical intervals produced by a vibrating string correspond to simple ratios of string lengths. The *octave*, for example, corresponds to a ratio of 2:1; the *fifth*, to 3:2, the *fourth*, to 4:3, and so on (the names octave, fifth, and fourth refer to the position of these intervals on the musical staff). Pythagoras took this as a sign that the entire universe—from the laws of musical harmony to the motion of the heavenly bodies—is governed by rational numbers. *Number Rules the Universe* became the Pythagorean motto.

But one day a member of the Pythagorean school by the name Hippasus made a startling discovery: the square root of 2—the number that when multiplied by itself results in 2—cannot be expressed as a ratio of two integers, no matter how much one tries.[1] You can approximate it as closely as you please by rational numbers, but you can never write it exactly as a ratio. For example, 14/10, 141/100, 1,414/1,000, and 14,142/10,000—or as decimals, 1.4, 1.41, 1.414, and 1.4142—are four rational approximations of $\sqrt{2}$, increasing progressively in accuracy. But to get the *exact* value of $\sqrt{2}$ would require us to write down an infinite, nonrepeating string of digits, and this cannot be expressed as a ratio of integers. Thus, $\sqrt{2}$ is an *irrational number* (a proof of the irrationality of $\sqrt{2}$ is given in the appendix).

The discovery that $\sqrt{2}$ is irrational shattered the Pythagorean belief in the rule of rational numbers, and it brought about a serious intellectual crisis. What to do with this new kind of number? Could it be represented geometrically? Take a square of unit side and draw its diagonal. By the Pythagorean theorem, this diagonal has a length equal to $\sqrt{2}$; but not being able to express it as a rational number, the Pythagoreans were forced to regard it as a purely geometric entity—in effect a line segment with an unde-

Plate 8. *This Is Not the Square Root of Two*

fined length! Their confusion can be seen from the double meaning of the phrase *irrational number*: a number that is not a ratio of two integers, but also an erratic number that defies rational behavior.

The crisis precipitated by this discovery had far-reaching consequences: it opened up a rift between the two major branches of mathematics, geometry and arithmetic (and, by extension, algebra), a rift that would impede the progress of mathematics for the next two thousand years. It was not until the invention of analytic geometry by Descartes and Fermat in the seventeenth century that the two branches were reunited.

You have probably seen posters of π or e (the base of natural logarithms) listing the first few thousand digits of their decimal expansion in row after row of monotonous figures. If you were wondering what are the first few digits of the decimal expansion of $\sqrt{2}$, plate 8 provides the answer in lively colors. We named it *This is Not the Square Root of 2*, paraphrasing René Magritte's (1898–1967) famous painting of a pipe, which he titled *Ceci n'est pas une pipe* ("this is not a pipe"). And, indeed, the long string of decimals in our illustration is *not* the square root of 2, just a close approximation of it.

NOTE:

1. More precisely, the Pythagoreans discovered that the numbers $\sqrt{2}$ and 1 are *incommensurable*—they do not have a common measure. That is, one cannot find two line segments of integer lengths m and n such that $n \cdot \sqrt{2} = m \cdot 1$. Had such line segments existed, it would mean that $\sqrt{2} = m/n$, a rational number.

$$9$$

A Repertoire of Means

Another subject of great interest to the Pythagoreans was how to find the average, or *mean*, of two positive numbers. At first thought this seems to be a trivial question. Let the numbers be a and b. Add them and divide by 2, getting $(a+b)/2$: you are done. Today, of course, we would compute this mean numerically; for example, the mean of 3 and 5 is $(3+5)/2 = \%_2 = 4$. The Greeks, however, thought of it in geometric terms: they regarded a and b as the lengths of two line segments, drawing them end to end along one line and finding the midpoint of the combined segment—all doable with straightedge and compass (see chapter 18).

This kind of mean, called the *arithmetic mean*, is just one of several possible means. Another often-used mean is the *geometric mean*, defined as $\sqrt{ab}$. Imagine you own a rectangular plot of land and you want to build on it a square-shaped house. Your plot, however, is rather long and narrow, so it doesn't provide enough space for your dream house. You therefore propose to your realtor to trade it off for a square-shaped plot of equal area. What should be the side of this square? If the length and width of the rectangle are a and b, its area is ab. Setting this equal to the area of a square of side x, we get $ab = x^2$, or $x = \sqrt{ab}$ —the geometric mean of a

and b. For example, the geometric mean of 3 and 5 is $\sqrt{3 \cdot 5} = \sqrt{15} \approx 3.87$, a shade less than their arithmetic mean.

Plate 9 depicts the geometric mean the way the Greeks looked at it—as a relation of areas. We think of a and b as the lengths of two line segments, and we draw them end to end along a straight line. We then draw a semicircle with this line as diameter and erect a perpendicular to it at the point where a and b are joined. The height of this perpendicular is $G = \sqrt{ab}$, the geometric mean of a and b (we will prove this in the next chapter). Therefore, $G^2 = ab$, so that the rectangle with sides a and b has the same area as a square with side G. The illustration shows a series of identical semicircles, all with the same diameter $a+b$ but with varying proportions $a : b$. Note that G reaches its maximum value when $a = b$, as shown in the central panel.

And there is a third mean, whose definition may at first seem a bit strange. A common situation encountered by pilots goes like this: an aircraft can fly at 500 mph in still air. But due to the presence of wind, this so-called airspeed will be different from the aircraft's *ground speed*. On a stretch between two cities, a pilot encounters a headwind of 50 mph, reducing the aircraft's ground speed to 450 mph. On

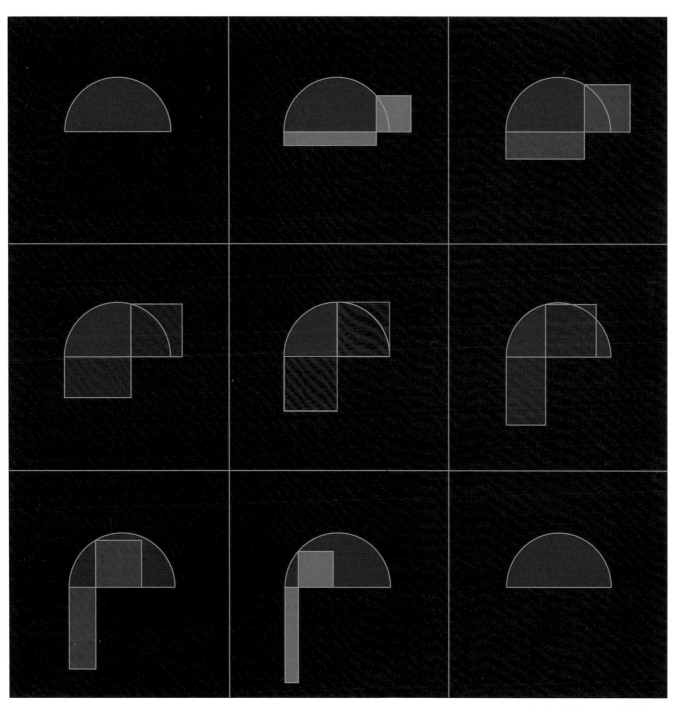

Plate 9. *Geometric Mean*

the return flight, the same wind now becomes a tail-wind, increasing the ground speed to 550 mph. At what speed would the aircraft have to fly *in still air* to complete the round trip in the same stretch of time as when the wind was blowing?

Your first impulse might be to give the answer as 500 mph, the aircraft's own airspeed and the arithmetic mean of 450 and 550. But first impulses can be wrong. Let the distance between the cities be d and the required speed, v. The time it takes the aircraft to complete the outbound stretch (flying against the headwind) is $d/450$, whereas the time of the return stretch is $d/550$. The total time is, therefore, $d/450 + d/550$. Setting this equal to the time it would take to complete the round trip at a constant ground speed v, we have

$$\frac{d}{450} + \frac{d}{550} = \frac{2d}{v}.$$

The distance d cancels out, leaving us with the equation

$$\frac{1}{450} + \frac{1}{550} = \frac{2}{v},$$

from which we get

$$\frac{1}{v} = \frac{1}{2}\left(\frac{1}{450} + \frac{1}{550}\right) = \frac{1}{2} \cdot \frac{450+550}{450 \cdot 550} = \frac{1,000}{495,000} = \frac{1}{495}$$

and, finally, $v = 495$ mph—slightly less than 500 mph. So the required speed is *not* the arithmetic mean of the two ground speeds, but a tad less.

This kind of mean, obtained by taking the *reciprocals* of the two numbers, finding their arithmetic mean, and then taking the reciprocal of the result, is called the *harmonic mean*. Putting this into the language of algebra, if H denotes the harmonic mean of a and b, we have

$$\frac{1}{H} = \frac{1}{2}\left(\frac{1}{a} + \frac{1}{b}\right) = \frac{a+b}{2ab}.$$

Taking the reciprocal of this, we get $H = 2ab/(a+b)$. The formulas $A = (a+b)/2$, $G = \sqrt{ab}$, and $H = 2ab/(a+b)$ comprise the three classical means of the Pythagoreans.

The names *arithmetic*, *geometric*, and *harmonic* for the three means derive from the three classical progressions with the same names: the arithmetic progression, in which there is a constant difference between successive terms; the geometric progression, which keeps a constant ratio, and the harmonic progression, whose terms are the reciprocals of those of an arithmetic progression. Here are three examples:

Arithmetic progression: 1, 2, 3, 4, 5, . . .

Geometric progression: 1, 2, 4, 8, 16, . . .

Harmonic progression: 1, ½, ⅓, ¼, ⅕, . . .

Each term in these progressions is, respectively, the arithmetic, geometric, and harmonic mean of the terms immediately preceding and following it. It may have been this connection that caught the Greeks' interest in these means.

10

More about Means

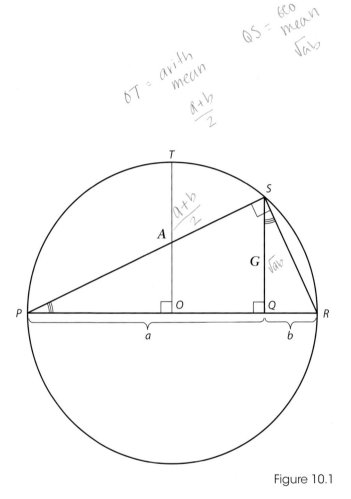

Figure 10.1

Turning again to the aircraft making its round trip between two cities, we found that the harmonic mean of the two ground speeds, 495 mph, was less—though just barely—than their arithmetic mean, 500 mph. This is not a coincidence. A well-known theorem says that of the three means, the harmonic mean is always the smallest, the arithmetic mean the largest, and the geometric mean somewhere in between. In the case of the aircraft, the geometric mean of the two speeds is $\sqrt{450 \cdot 550} = 497.49$ mph, rounded to two places, so we have $H < G < A$. This double inequality can actually be made more general: the less-than signs become equal signs if and only if the two numbers whose mean we are seeking are equal. That is,

$$H \leq G \leq A,$$

with equality if and only if $a = b$. This statement is known as the *arithmetic-geometric-harmonic mean inequality*.

To see why this is so, let us use a geometric construction, in the spirit of the Greeks. Suppose we are given two line segments of lengths a and b. Place them end to end along a line, with $a = \overline{PQ}$ and $b = \overline{QR}$ (figure 10.1). Bisect $\overline{PR}$ and call its midpoint O. We have $\overline{OP} = (a+b)/2 = A$, the arithmetic mean of a and b.

Now draw a semicircle with center at O and diameter $\overline{PR}$ (using the same notation as before; again see figure 10.1). At Q erect a perpendicular to $\overline{PR}$, meeting the circle at S. By Thales's theorem (see chapter 1), $\angle PSR = 90°$ and, therefore, triangles PQS and SQR

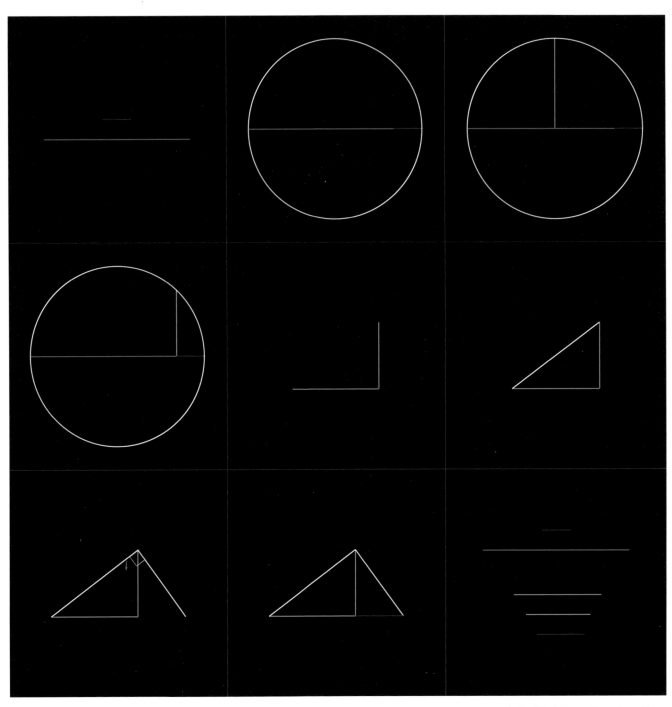

Plate 10. *Mean Constructions*

are similar, having a common right angle at Q and equal angles QPS and QSR. Thus, $\overline{PQ}/\overline{QS} = \overline{SQ}/\overline{QR}$, or $\overline{QS}^2 = \overline{PQ}\cdot\overline{QR} = a\cdot b$ (we consider all line segments to be nondirectional, so $\overline{QS} = \overline{SQ}$). This gives us $\overline{QS} = \sqrt{ab} = \boldsymbol{G}$, the geometric mean of a and b.

To show that the geometric mean of a and b cannot be larger than their arithmetic mean, we refer again to figure 10.1. Draw a line segment perpendicular to $\overline{PR}$ at O, meeting the circle at T. Because $\overline{OP}$ is the radius, we have $\overline{OT} = \overline{OP} = \boldsymbol{A}$. It is then clear from our figure that $\overline{QS} \leq \overline{OT}$, that is, $\boldsymbol{G} \leq \boldsymbol{A}$. Furthermore, the two means are equal if and only if points S and T coincide, in which case $\overline{PQ} = \overline{QR}$, that is, $a=b$.

What about the harmonic mean? Can it be constructed as easily as its arithmetic and geometric counterparts? The answer is yes, but in order to do so we must first prove a rather surprising result: if we multiply together the expressions $\boldsymbol{A}=(a+b)/2$ and $\boldsymbol{H}=2ab/(a+b)$, we get $\boldsymbol{AH}=\left[(a+b)/2\right]\cdot\left[2ab/(a+b)\right]=ab=\boldsymbol{G}^2$, or $\boldsymbol{G}=\sqrt{\boldsymbol{AH}}$: *the geometric mean of a and b is also the geometric mean of $\boldsymbol{A}$ and $\boldsymbol{H}$.* This result is the key to the construction of $\boldsymbol{H}$. For, if we rewrite the equation $\boldsymbol{G}^2=\boldsymbol{AH}$ as a proportion, $\boldsymbol{A}/\boldsymbol{G}=\boldsymbol{G}/\boldsymbol{H}$, we see that $\boldsymbol{A}$ and $\boldsymbol{H}$ play the same role vis-à-vis $\boldsymbol{G}$ as did a and b in our construction of the geometric mean.

Thus, having already found $\boldsymbol{A}$ and $\boldsymbol{G}$ from figure 10.1, construct right triangle $P'S'Q'$ with $\overline{P'Q'}=\boldsymbol{A}$ and $\overline{Q'S'}=\boldsymbol{G}$ (figure 10.2). The perpendicular from S' to

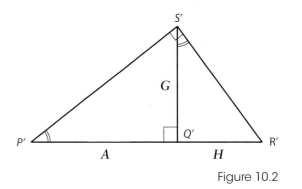

Figure 10.2

$\overline{P'S'}$ meets the extension of $\overline{P'Q'}$ at R', with $\overline{Q'R'}=\boldsymbol{H}$. And since we already know that $\boldsymbol{G}\leq\boldsymbol{A}$, it follows from the similarity of the triangles $P'Q'S'$ and $S'Q'R'$ that $\boldsymbol{H}\leq\boldsymbol{G}$. Combining the two inequalities, we get $\boldsymbol{H}\leq\boldsymbol{G}\leq\boldsymbol{A}$, the arithmetic-geometric-harmonic mean inequality.

It is truly remarkable that the circle— perhaps the simplest of all geometric constructs—holds within it so many hidden features waiting to be discovered by a keen observer. No wonder the Greeks held the circle in such high esteem.

Plate 10, *Mean Constructions* (no pun intended!), is a color-coded guide showing how to construct all three means from two line segments of given lengths (shown in red and blue). The arithmetic, geometric, and harmonic means are colored in green, yellow, and purple, respectively, while all auxiliary elements are in white.

11

Two Theorems from Euclid

Theorem 35 of the third book of the *Elements* says, *If in a circle two straight lines cut one another, the rectangle contained by the segments of the one is equal to the rectangle contained by the segments of the other.*

To make sense of this enigmatic statement, we must understand that the Greeks always used geometric language to describe operations that nowadays would be stated in algebraic terms. Thus, "the rectangle contained by" is code for "the product of" [the sides of the rectangle]—in other words, the area of the rectangle. Translated into modern language, the theorem says: if through a point P inside a circle we draw a line that cuts the circle at points A and B, the product $\overline{PA} \times \overline{PB}$ is constant; it has the same value for all lines through P (see figure 11.1, where $\overline{PA} \times \overline{PB} = \overline{PC} \times \overline{PD}$).

This is followed by theorem 36: *If a point be taken outside a circle and from it there fall on the circle two straight lines, and if one of them cut the circle and the other touch it, the rectangle contained by the whole of the straight line which cuts the circle and the straight line intercepted on it outside between the point and the convex circumference will be equal to the square on the tangent.*

Again, behind this seemingly convoluted language is the statement: if from a point P outside a

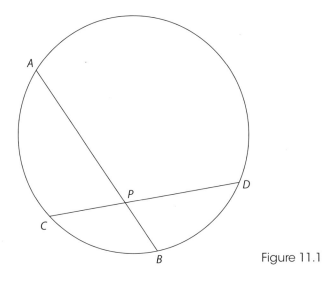

Figure 11.1

circle we draw a line that cuts the circle at points A and B, the product $\overline{PA} \times \overline{PB}$ is constant for all possible lines through P and is equal to the square of the length of the tangent line from P to the circle (figure 11.2, where $\overline{PA} \times \overline{PB} = \overline{PC} \times \overline{PD} = \overline{PT}^2$). Note that the phrase *the square on the tangent* actually means the *area* of a square whose side equals the length of the tangent line.

Plate 11. *Circles and Rectangles*

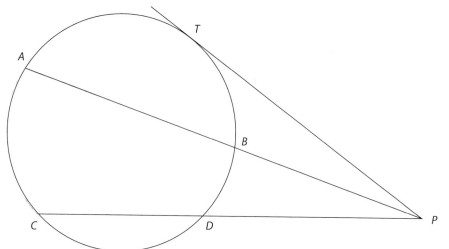

Figure 11.2

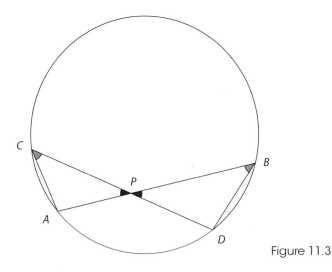

Figure 11.3

gles $\angle ACD$ and $\angle DBA$ subtend the same arc, $\overset{\frown}{AD}$, on the circumference (Euclid III, 21). Thus, triangles *PAC* and *PDB* are similar, having two pairs (and, therefore, three) of equal angles. Consequently, $\overline{PA}/\overline{PC} = \overline{PD}/\overline{PB}$, or $\overline{PA} \times \overline{PB} = \overline{PC} \times \overline{PD}$, which means that the product $\overline{PA} \times \overline{PB}$ is constant for all possible lines through *P*.

The proof of theorem 36 follows similar lines, and we leave it to the reader to work out the details. The additional statement, that the product $\overline{PA} \times \overline{PB}$ is equal to the square of the tangent line from *P* to the circle, can be thought of as a limiting case when point *A* approaches point *B*, ultimately to coincide with the point of tangency *T*. We then have $\overline{PA} \times \overline{PB} = \overline{PT} \times \overline{PT} = \overline{PT}^2$. We should note, however, that the Greeks would not have accepted such a limiting argument into their reasoning because it subtly involves the notions of motion and

The proof of theorem 35 is quite simple. In figure 11.3, *P* is a point inside the circle, and $\overline{AB}$ and $\overline{CD}$ are two chords passing through *P*. We have $\angle APC = \angle BPD$ and $\angle ACD = \angle DBA$, the latter equality because an-

continuity, two concepts that Euclid regarded as physical in nature and thus foreign to pure mathematical thinking.

Our illustration (plate 11) shows four identical circles, each with a pair of chords intercepting at P. The two upper circles (with P outside each circle) illustrate theorem 36 and the two lower circles (P inside), theorem 35. In each case, $\overline{PR} \times \overline{PS}$ is represented by a rectangle of sides $\overline{PR}$ and $\overline{PS}$ (shown in orange), and since this product is constant for all possible chords through P, the two rectangles on either side of each circle are equal in area.

12

Different, yet the Same

The two theorems we just met—numbers 35 and 36 in book III of Euclid—sound strikingly similar: both are about a circle, a point P, and a line through P that cuts the circle at points R and S. The two theorems state that the product $\overline{PR} \times \overline{PS}$ remains constant for all possible lines through P. Yet there is a difference: in theorem 35 P is inside the circle, while in theorem 36 it is outside. And theorem 36 has the additional result that $\overline{PR} \times \overline{PS} = \overline{PT}^2$, where $\overline{PT}$ is the length of the tangent line from P to the circle. So despite their similarity, they are two distinct statements, and Euclid was careful to list them as separate theorems, one following the other.

And yet a closer inspection will show that the two theorems are one and the same after all. For when we say that a point P is given with respect to a circle, we imply only that the distance $\overline{OP}$ from P to the center of the circle O is given, and nothing else. The actual position of P relative to the circle (that is, its direction from O) is immaterial. And since all points P with the same distance $\overline{OP}$ from O describe a second circle concentric to the given circle and having the radius $\overline{OP}$, we conclude that $\overline{PR} \times \overline{PS}$ has the same value for all points on this second, "implied" circle.

Let us denote the two circles by (O, p) and (O, r), where O is their common center and $p = \overline{OP}$ and $r = \overline{OR}$ are their radii (see figure 12.1, where (O, p) is marked in red and (O, r) in blue). We notice that the two circles can be interchanged by simply switching the roles of P and R and of S and Q. In figure 12.1(a), P is outside (O, r) and we have $\overline{PR} \times \overline{PS} = \overline{PT}^2$; in (b), P is inside (O, r) and we have $\overline{RP} \times \overline{RQ} = \overline{RT}^2$. But note that $\overline{RQ} = \overline{PS}$, because P and Q are symmetrically positioned with respect to O (as are R and S); consequently, we can rewrite the last equation as $\overline{PR} \times \overline{PS} = \overline{PT}^2$. But this is exactly the same equation as in case (a)! We can, therefore, combine theorems 35 and 36 into a single statement:

If through a point P a line is drawn that cuts a circle (O, r) at points R and S, the product $\overline{PR} \times \overline{PS}$ is constant for all possible lines through P. If P is outside (O, r), this product is the square of the tangent from P to (O, r); if P is inside (O, r), it is the square of the tangent from R to (O, p).

Plate 12 illustrates the complete symmetry between the two theorems. The top two figures illustrate theorem 36, the bottom two, theorem 35. The "given" and "implied" circles are marked in blue and red, respectively; the yellow rectangles and squares

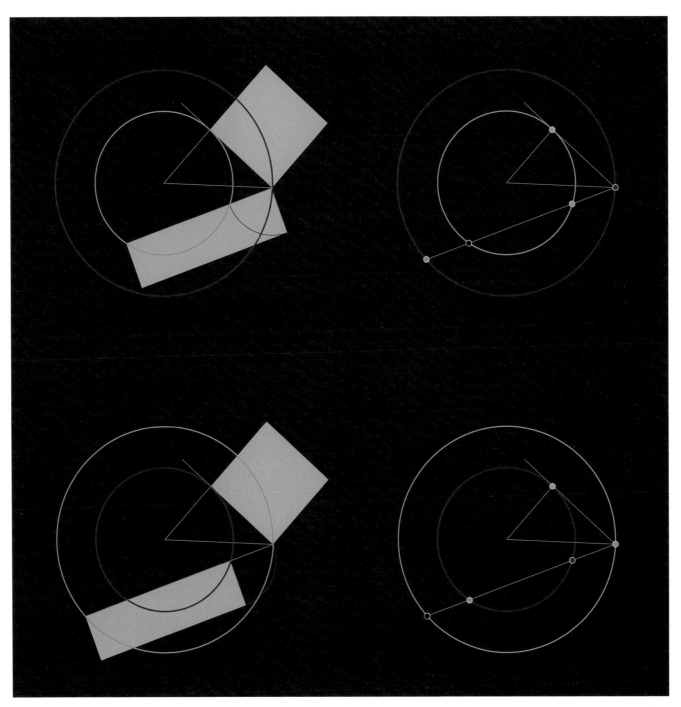

Plate 12. *Euclid I, 35 and 36*

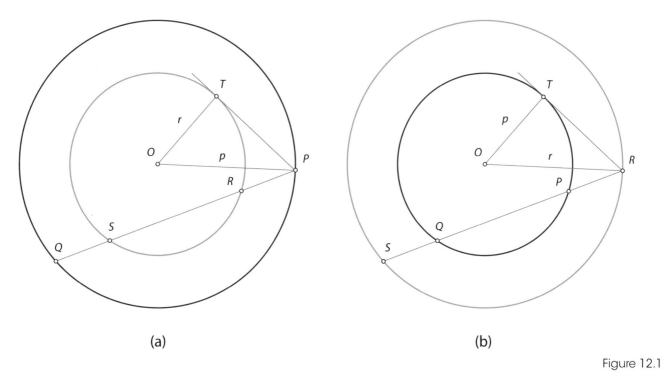

(a)   (b)

Figure 12.1

represent the product $\overline{PR} \times \overline{PS}$ and the quantities $\overline{PT}^2$ (in the top figure) and $\overline{RT}^2$ (bottom figure). The two configurations are exactly the same, except that the given and implied circles reverse their roles.

The French mathematician Lazare Carnot (1753–1823) achieved this unification by regarding all line segments as *directed* quantities that can assume positive or negative values (for example, $\overline{PR} = -\overline{RP}$). This, however, would not have sat well with Euclid, because the Greeks did not recognize negative quantities. Why, then, didn't Euclid think of combining the two theorems the way we did here (that is, by regarding all line segments as nondirected quanti-

ties)? Surely he must have noticed the similarity between the two, but the idea of generalizing several particular cases into a broader, sweeping statement was foreign to the Greeks. To Euclid, each case represented a separate theorem, standing firmly on its own. Generalizations had to wait for future generations of mathematicians.[1]

NOTE:

1. This chapter is based on an article by Maor published in *The Mathematics Teacher* (May 1979, pp. 363–367).

13

One Theorem, Three Proofs

Theorem 13 of Book VI of Euclid tells us how to find the geometric mean (the *mean proportion,* as the Greeks called it) of two line segments. In essence, it says that in a right triangle, the altitude h divides the hypotenuse into two segments m and n such that $h/m = n/h$. From this it follows that $h^2 = mn$, so that h is the geometric mean of m and n. Plate 13 illustrates this for $m = 9$, $n = 4$, and $h = 6$.

We offer here three quite different proofs, with the question in mind, which of them is the simplest? Let the triangle be ABC, with the right angle at C (figure 13.1). From C drop the altitude $\overline{CD} = h$ to the hypotenuse, dividing it into segments $\overline{AD} = m$ and $\overline{DB} = n$. The first proof—the one found in most geometry textbooks—relies on the similarity of triangles ADC and CDB, since they share a right angle at D and equal angles $\angle DAC$ and $\angle DCB$. Thus, $\overline{AD}/\overline{DC} = \overline{CD}/\overline{DB}$, or $m/h = h/n$, or $h^2 = mn$ (as before, all line segments are nondirectional, so that $\overline{CD} = \overline{DC}$).

The second proof is a direct consequence of Thales's theorem (or rather its converse; see chapter 1, note 1) and Euclid III 35 (chapter 11). Inscribe triangle ABC in a circle with diameter AB, as in figure

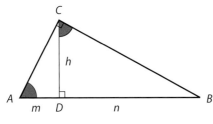

Figure 13.1

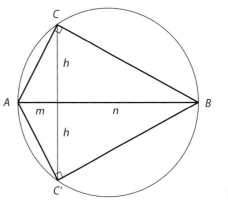

Figure 13.2

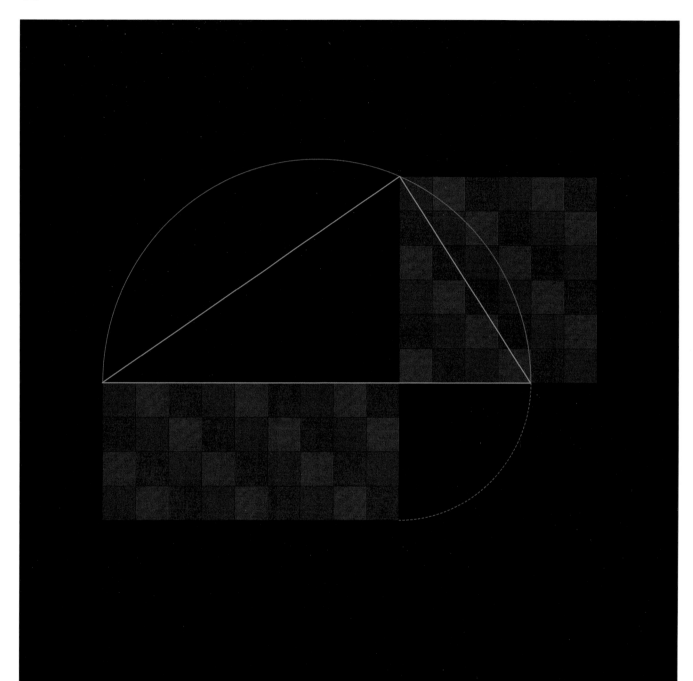

Plate 13. $6^2 = 9 \times 4$

13.2. This circle passes through C and through its mirror image C' when reflected in the diameter. Hence $m \cdot n = h \cdot h = h^2$—QED.

The third proof relies on a comparison of areas. We turn triangle ADC counterclockwise through 90° about D to get the skewed butterfly-like figure $BCDA'C'D$ (figure 13.3). Note that angles $\angle C'A'D$ and $\angle BCD$ are equal, and therefore lines BC and $A'C'$ are parallel, with CA' acting as a transversal (Euclid I 27). We now construct square $CD'C'D$ with area h^2 and rectangle $A'B'BD$ with area mn, and divide each into two halves by diagonals $C'C$ and $A'B$. Triangles $C'A'B$ and $C'A'C$ have the same area, since their vertices B and C lie on a line parallel to the base $A'C'$ (Euclid I 38). From each of these triangles subtract triangle $A'DC'$, resulting in triangles $C'DC$ and $A'DB$ having the same area. But triangle $C'DC$ has half the area of square $CD'C'D$, and triangle $A'DB$ has half the area of rectangle $A'B'BD$. Thus, $h^2 = mn$.

We return now to the question we posed at the beginning of this chapter: which of the three proofs is the simplest? Judging by their length, the second proof is the obvious winner, requiring just four lines of explanatory text. But length is only one criterion of what constitutes a simple proof. Another criterion—and arguably a more important one—is how many previously established propositions the theorem in question is directly based upon. And this puts the third proof up front: it rests on just two earlier propositions—Euclid I 27 and I 38. Moreover, it is in line with the Greek interpretation of a product as the area of a rectangle, and in this sense it echoes Euclid's famous proof of the Pythagorean theorem (Euclid I 47).

Of course, even this criterion is not entirely foolproof, as each of the two theorems on which the last

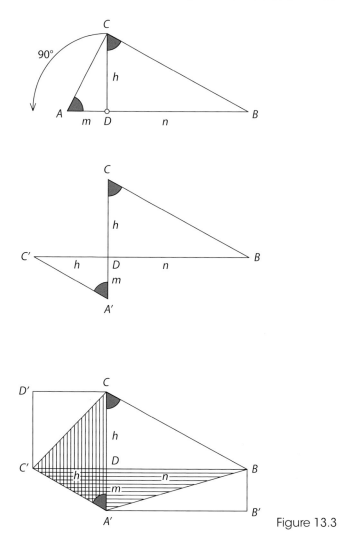

Figure 13.3

proof relies is itself resting on several earlier theorems. The complete ancestry of a theorem may be long and arduous, making any absolute judgment of which proof is the simplest almost impossible. Ultimately, simplicity is in the eyes of the beholder.

14

The Prime Numbers

The prime numbers have always enjoyed a special status among the integers. A prime number, or *prime*, for short, is an integer greater than 1 that can be divided only by itself and 1. The first 10 primes are 2, 3, 5, 7, 11, 13, 17, 19, 23, and 29. The smallest—and the only even prime—is 2. The largest, as of this writing, is $2^{57,885,161} - 1$, a gargantuan 17,425,170-digit number that would fill some 2,500 pages if printed.[1] An integer greater than 1 that is not prime is called *composite*. The number 1 is considered neither prime nor composite.

The importance of the primes in number theory—and by extension, in much of mathematics—comes from the fact that *every positive integer can be factored into primes in one and only one way*. For example, 12 can be factored into 3 and 4, so $12 = 3 \times 4$. But 4 itself can be factored into 2 and 2, so we end up with $12 = 3 \times 2 \times 2$. We could have started by factoring 12 into 2 and 6; but $6 = 2 \times 3$, so we get $12 = 2 \times 2 \times 3$. Except for their order, we end up with the same set of primes. This fact, known as *the fundamental theorem of arithmetic*, is true for every positive integer greater than 1 (although for large numbers the factorization process may take a very long time and may even be impossible to achieve in practice). The primes are,

therefore, considered the building blocks of all integers, playing a role somewhat analogous to that of the chemical elements in the periodic table. In fact, so fundamental are the primes to mathematics that it has been proposed to use them as a means of communication with extraterrestrial beings, if and when we find them.

Plate 14 shows a row of dots representing the number 17. As much as you may try, you cannot arrange these dots in a rectangular array: there will always be at least one left over. That, of course, is because 17 is prime—it is not the product of any smaller integers except itself and 1.

Primes of the form $2^n - 1$, like the prime $2^{57,885,161} - 1$ just mentioned, are known as *Mersenne primes*, after the Minimite friar and freelance mathematician Marin Mersenne (1588–1648). For $2^n - 1$ to be prime, n itself must be prime, but the converse is false: $2^{11} - 1 = 2,047 = 23 \times 89$. The first five Mersenne primes are $2^2 - 1 = 3$, $2^3 - 1 = 7$, $2^5 - 1 = 31$, $2^7 - 1 = 127$, and $2^{13} - 1 = 8,191$. To date, only 48 Mersenne primes are known; the largest, discovered in 2013, is the number quoted earlier. It is not known how many Mersenne primes exist—or even if their number is finite or infinite.

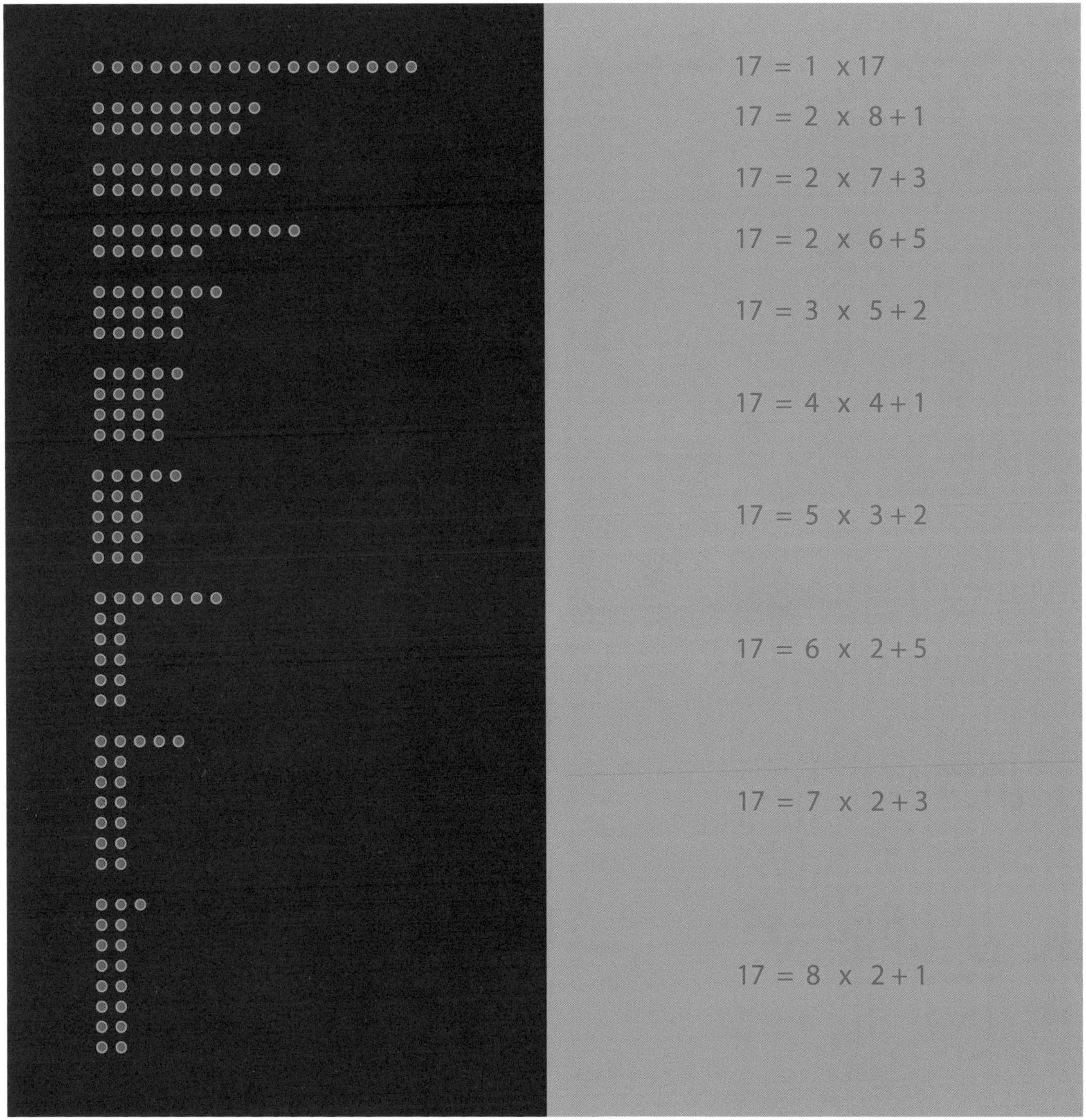

$$17 = 1 \times 17$$

$$17 = 2 \times 8 + 1$$

$$17 = 2 \times 7 + 3$$

$$17 = 2 \times 6 + 5$$

$$17 = 3 \times 5 + 2$$

$$17 = 4 \times 4 + 1$$

$$17 = 5 \times 3 + 2$$

$$17 = 6 \times 2 + 5$$

$$17 = 7 \times 2 + 3$$

$$17 = 8 \times 2 + 1$$

Plate 14. *17 Is Prime*

In chapter 4 we mentioned that if $2^n - 1$ is prime, then $2^{n-1} \cdot (2^n - 1)$ is perfect. Consequently, each newly discovered Mersenne prime automatically yields a new perfect number. Since the largest prime to date, $2^{57,885,161} - 1$, is the 48th known Mersenne prime, it follows that $2^{57,885,160} \cdot (2^{57,885,161} - 1)$ is the 48th known perfect number.

Many questions about the primes are still unanswered, adding to the aura of mystery that has always surrounded these numbers. Two questions immediately come to mind: how many primes are there, and what is their law of distribution among the integers? The first question was answered by Euclid around 300 BCE. In a classic proof that became a model of simplicity, Euclid showed that there is no end to the primes: their number is infinite. Although the primes, on the average, thin out as we go to higher numbers, there is no last prime beyond which all integers are composite. We give Euclid's proof in the appendix.

The second question—the distribution of primes among the integers—has intrigued mathematicians for centuries. In contrast to the elements of the periodic table, the primes do not seem to follow any recognizable pattern; in fact, all attempts to find a formula that would generate all primes have so far failed. We do know, however, something about their *average* distribution. In 1792, when he was just 15 years old, Carl Friedrich Gauss (1777–1855) examined a table of primes and announced that the number of primes below a given integer n increases approximately as $n/\ln n$, where ln stands for *natural logarithm* (logarithm to the base $e = 2.71828\ldots$; we'll

Figure 14.1

come back to this number in chapter 33). Moreover, the approximation gets better with increasing n and approaches $n/\ln n$ as $n \to \infty$. It was only in 1896, more than a hundred years after Gauss's announcement, that Jacques Solomon Hadamard of France and Charles de la Valleé-Poussin of Belgium independently proved his conjecture. It is known as the *prime number theorem*.

In years past, the discovery of a new prime often made the news. In 1963, the University of Illinois issued a special cancellation stamp to commemorate their discovery of the largest prime then known: $2^{11,213} - 1$ (figure 14.1). This 3,376-digit number—the twenty-third Mersenne prime—was huge at the time of its discovery but is dwarfed by the primes that have been discovered since. But rest assured, still larger primes are lurking just around the corner, waiting to be discovered by today's powerful supercomputers—or even home computers.

NOTE:

1. Source: *The Largest Known Primes—A Summary*, on the Web at http://primes.utm.edu/largest.html, 2013.

15

15

Two Prime Mysteries

Among the many unsettled questions about the primes, two stand out for their deceptive simplicity. Even a cursory glance at a table of primes will reveal the abundance of pairs of primes of the form p and $p+2$: 3 and 5, 5 and 7, 11 and 13, ... , 101 and 103, and so on. One can find these *twin primes* even among very large numbers: 29,879 and 29,881, 140,737,488,353,699 and 140,737,488,353,701. At the time of writing, the largest known twin pair is $3{,}756{,}801{,}695{,}685 \cdot 2^{666{,}669} \pm 1$, each having 200,700 digits.[1] How many twin primes are there? Mathematicians are nearly unanimous in their belief that there exist infinitely many of them, just like the primes themselves. But belief counts for very little in mathematics; what counts is proof, and thus far no one has been able to prove this *twin prime conjecture*.

The second unanswered question was first raised by Christian Goldbach (1690–1764), a German mathematics teacher and, later, a diplomat in the service of the Russian czar. In 1742 he wrote a letter to Leonhard Euler, the foremost mathematician of the time, in which he reported on a startling observation: every even number greater than 2 he had examined was the sum of two primes (sometimes in more than one way). For example, $4=2+2$, $6=3+3$, $8=5+3$, $10=5+5=7+3$, and so on. (Note that for *odd* numbers this is false: 11 cannot be written as the sum of two primes). Goldbach conjectured that this is true for *all* even numbers. He tried to prove it but failed, so he posed the question to Euler. Euler, whose mind was occupied with more pressing mathematical problems, shelved Goldbach's letter; it was discovered only after Euler's death in 1783 among his enormous volume of correspondence. Despite numerous attempts to prove the conjecture or find a counterexample, Goldbach's conjecture remains unsettled.

Until a few decades ago, the primes were considered the ultimate object of pure mathematics, existing in the ethereal universe of number theory and devoid of any practical applications. But that has recently changed: since about 1980 the primes have played a central role in encoding financial transactions and confidential communications to ensure the security of the transmitted information over the Internet. Thus, it is always possible that a once obscure and abstract subject may suddenly find a practical application in the real world.

Plate 15.1, *Prime and Prime Again*, shows a curious number sequence: start with the top eight-digit

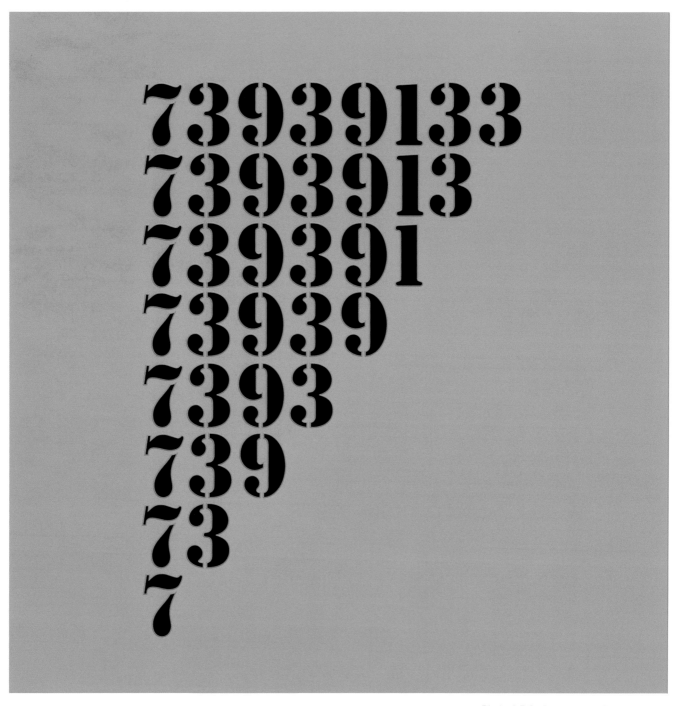

Plate 15.1. *Prime and Prime Again*

Plate 15.2. *Homage to Martin Gardner*

number and keep peeling off the last digits one by one, until only 7 is left. For no apparent reason, each number in this sequence is a prime. Plate 15.2 is a quote from the eminent popularizer of mathematics Martin Gardner (1914–2010), whose many articles on the primes made them a household name. Notice the dots surrounding the text, starting with the arrow at the top: if you look carefully, you'll discover that the second, third, fifth, seventh, eleventh, . . . dots are marked in red—the prime numbers

in order. If aliens from another planet would look at this picture, they would most likely not be able to read its textual message, but they just might discover the hidden primes around it—and perhaps respond in kind.

NOTE:

1. Source: *The Largest Known Primes—A Summary*, on the Web at http://primes.utm.edu/largest.html, 2013.

16

$0.999\ldots = ?$

When I ask beginning mathematics students, "is 0.999... *exactly* equal to 1, or only approximately so?" their responses are usually split evenly, but occasionally the majority will vote for the second option. Well, let's see:

$$x = 0.999\ldots$$
$$10x = 9.999\ldots.$$

Subtract the first equation from the second:

$$9x = 9$$
$$x = 1\,.$$

Surprising?... If this simple question can cause such disagreement today, how much more so in ancient times, when the idea of anything going on to infinity was so confusing that it was avoided entirely, shunned by the Greeks as *horror infiniti*, the horror of the infinite.

To make the point, the fourth-century BCE philosopher Zeno of Elea came up with four paradoxes—he called them "arguments"—meant to show that mathematicians were unable to deal with infinity. In one of these paradoxes Zeno purports to show that motion is impossible. Imagine an athlete about to run a 1-mile stretch. To do so, the runner would first have to cross the halfway point, then half of

what's still left (a quarter of the total distance), then half of that (an eighth), and so on ad infinitum (see figure 16.1). "Impossible!" said Zeno. No human can exhaust an infinite number of steps, so there will always be a tiny distance still left to be covered. The runner will be unable to reach the finish line; in fact, using the same argument, *any* kind of motion is impossible. Yet we do it all the time, giving no thought at all to the process.

At the heart of the runner's paradox is the infinite series

$$\frac{1}{2}+\frac{1}{4}+\frac{1}{8}+\frac{1}{16}+\cdots.$$

Such a series, in which each term has a fixed ratio to its predecessor, is called a *geometric series* (for no particular connection to geometry). In the general case, a geometric series can be written as

$$a+ar+ar^2+\cdots,$$

where a is the initial term and r, the common ratio. If r is less than 1 in absolute value (that is, $-1<r<1$), the series will *converge*—it will reach a definite limit as the number of terms keeps growing. That is to say, by adding more and more terms, the sum will get ever closer to its limiting value, making the re-

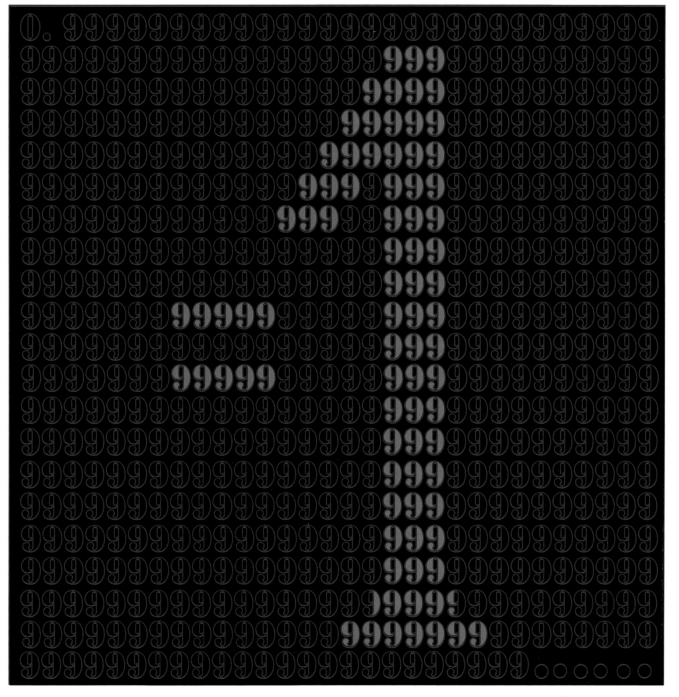

Plate 16. *0.999... = 1*

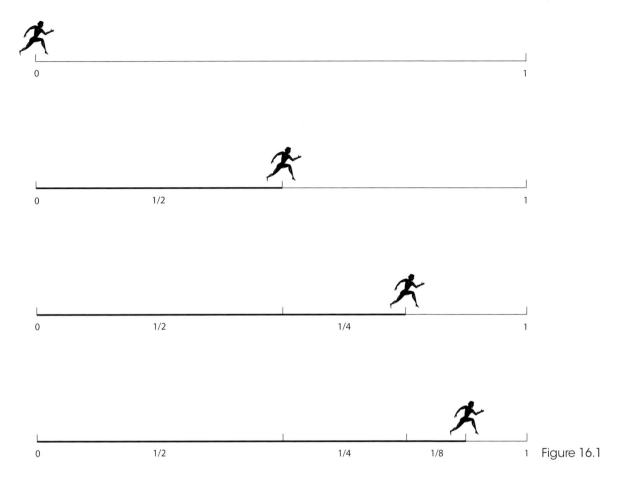

Figure 16.1

maining difference as small as we please. In the case of the runner, we have

$$\frac{1}{2} = 0.5,$$

$$\frac{1}{2} + \frac{1}{4} = \frac{3}{4} = 0.75,$$

$$\frac{1}{2} + \frac{1}{4} + \frac{1}{8} = \frac{7}{8} = 0.875,$$

$$\frac{1}{2} + \frac{1}{4} + \frac{1}{8} + \frac{1}{16} = \frac{15}{16} = 0.9375,$$

and so on. The sum seems to be getting ever closer to 1, the limiting value of the series.

Of course, adding the first few terms of our series does not *prove* that it converges to 1—or even that a limiting sum exists in the first place. Returning to the general case, it is not difficult to show that as long as $-1 < r < 1$, the limiting sum is $a/(1-r)$ (we give the proof in the appendix). For the runner's paradox we have $a = \frac{1}{2}$ and $r = \frac{1}{2}$, so the sum is $(\frac{1}{2})/(1 - \frac{1}{2}) = 1$, and the runner will reach the finish line just fine.

Going back to the question we posed at the beginning of this chapter—is 0.999… *exactly* equal to 1, or only approximately so?—we note that the repeating decimal 0.999… is actually a geometric series, $9/10 + 9/100 + 9/1,000 + \cdots$, whose initial term is $a = 9/10$ and whose common ratio is $r = 1/10$. Since this common ratio satisfies the condition $-1 < r < 1$, the series will converge to the limit $(9/10)/(1 - 1/10) = (9/10)/(9/10) = 1$, settling the issue once and for all.

Zeno's paradoxes caused a stir in the mathematical community that lasted well over two thousand years. Despite volumes of arguments, mostly philosophical or religious, no one was able to offer a convincing refutation of the paradoxes. And no wonder: to explain them, one must first accept the existence of infinity as a mathematical reality, a mental leap that even nineteenth-century mathematicians were not quite ready to take. It took the insight of a relatively unknown genius by the name Georg Cantor to take this crucial step, and in doing so he revolutionized our understanding of infinity. We will take a closer look at Cantor's ideas in chapter 51. However, if you are still wondering how a string of nines can make up a 1, plate 16, *0.999… = 1* will provide an answer—albeit a whimsical one.

17

Eleven

There is a parody about a mathematician who tries to prove that all numbers (here meaning positive integers) are interesting. Assume not. The number 1 is certainly interesting, being the generator of all numbers. So is 2, the first even integer and the only even prime. Three, being the sum of 1 and 2, makes it interesting as well. What about 4? We have $4 = 2 + 2 = 2 \times 2 = 2^2$: no doubt about it, 4 is definitely interesting. And so it goes, until we arrive at the first uninteresting number. But this, of course, makes it interesting! Thus, all numbers are interesting—QED.

If we had to choose an uninteresting number, 11 would certainly be a candidate. Tucked unceremoniously between its two more famous neighbors, 10 and 12, it seems to lack any defining characteristics: it is not a member of any immediately recognizable number sequence,[1] nor is it a perfect square or a sum of two squares. In mythology, too, 11 seems to have acquired a negative reputation: in ancient Rome, an assembly of 11 men was charged with apprehending criminals and bringing them to justice—a precursor of our modern jury system. The sixteenth-century numerologist Petrus Bungus deemed 11 as having "no connection to divine things, no ladder reaching up to things above, nor any merit."[2] Even nature seems to shun 11: flowers with 2, 3, 4, 5 and 6 petals are very common, but not with 11. Nor does it play a role in the inorganic world of crystals and minerals. In Peter Stevens's exhaustive *Handbook of Regular Patterns* (see the bibliography), with its hundreds of designs taken from all aspects of art and nature, not a single pattern is based on 11.

Still, the excluded number has its advocates: as told in Genesis 37:9, 11 stars appear in Joseph's dream, together with the Sun and Moon—perhaps a reference to the 11 constellations of the zodiac visible on any given night, the twelfth being hidden by the Sun. The Susan B. Anthony dollar coin, still in circulation but rarely used, is framed by an 11-sided regular polygon (figure 17.1). And of course, 11 is the number of players in a football team—American football as well as soccer.

Eleven does have some mathematical claim to fame: it is the second *repunit*—a number, all of whose digits are 1. These numbers are denoted by R_n, where n is the number of 1s. So $R_1 = 1 = (10 - 1)/9$, $R_2 = 11 = (100 - 1)/9$, and, in general, $R_n = (10^n - 1)/9$. These repunits have some remarkable properties: not only are all repunits *palindromes* (numbers that are the same whether read forward or backward), but so are their squares, up to R_9:

Plate 17. *Celtic Motif 1*

Figure 17.1

$$R_1^2 = \qquad 1$$
$$R_2^2 = \qquad 121$$
$$R_3^2 = \qquad 12321$$
$$R_4^2 = \qquad 1234321$$
$$\cdots$$
$$R_9^2 = 12345678987654321$$

Eleven is one of a small number of primes that have simple divisibility rules. Take the number 1,529 and alternately add and subtract its digits, going from left to right: $1-5+2-9=-11$. Since the result is divisible by 11, so is the number itself (indeed, $1,529 = 11 \times 139$). You can check this for as many numbers as you wish: it always works.[3] Repunits in general follow some simple divisibility rules: No repunit is divisible by 2 or 5; it is divisible by 3 if and only if n is a multiple of 3; by 7 and by 13 if and only if n is a multiple of 6; and by 11 if and only if n is even. Among composite repunits, R_{38} has a particularly interesting prime factorization: $11 \times 909,090,909,090,909,091 \times$

1,111,111,111,111,111,111, of which the first and third factors are themselves repunits, R_2 and R_{19}, while the second factor, except for the trailing 1, has a similar digit structure as that of $1/11 = 0.\overline{09}$.

As with Mersenne numbers (see page 42), R_n can be prime only if n is prime, but the converse is false: $R_3 = 111 = 3 \times 37$. As of this writing, only five prime repunits are known: R_2, R_{19}, R_{23}, R_{317}, and R_{1031}, the last discovered in 1986 by Hugh C. Williams and Harvey Dubner. In addition, there are several "probable primes" whose primality has still to be confirmed. And like the Mersenne primes, it is unknown how many repunit primes exist—or even if their number is finite or infinite.[4]

Our illustration (plate 17) shows an intriguing lace pattern winding its way around 11 dots arranged in three rows; it is based on an old Celtic motif. We hope this excursion into an "unpopular" number will encourage the reader to search for it in other places and be rewarded with discovering the unexpected.[5]

NOTES:

1. It is, however, the fifth member of the *Lucas numbers*—a Fibonacci-like sequence (see chapter 20) that starts with 1 and 3: 1, 3, 4, 7, 11, 18, 29, ...

2. Annemarie Schimmel. *The Mystery of Numbers* (New York: Oxford University Press, 1993), p. 189.

3. Sometimes the result may be 0, as, for example, with 187. Since 0 is divisible by 11 ($0 = 11 \times 0$), so is 187; indeed, $187 = 11 \times 17$.

4. Source: *Wolfram MathWorld,* on the Web at http://mathworld.wolfram.com/Repunit.html, 2013.

5. This chapter is based on an article by Maor in the journal *Mathematics Teaching in the Middle School* (January 2002).

18

Euclidean Constructions

According to tradition, it was Plato (ca. 427–347 BCE) who decreed that all geometric constructions should be done with a straightedge (an unmarked ruler) and compass alone. Of course, there is nothing intrinsically special about these tools, except perhaps their simplicity (you can still get them for a dollar or two at any drugstore), but Plato made their use into an art. Hundreds of constructions can be done with them, from very basic drawings to highly complex designs. Indeed, straightedge and compass constructions became so fundamental to geometry that Euclid incorporated them in his *Elements* from the very beginning. Proposition 1 of Book I—the very first of the 465 theorems in his *Elements*—shows us how to construct an equilateral triangle when its side is given. It is the same construction that generations of students have learned in their geometry class:

Let the given side be line segment $\overline{AB}$ (figure 18.1). With A as center, draw circle BCD. With B as center, draw circle ACE. The two circles meet at C. Now join C with A and with B. We have $\overline{AB} = \overline{AC}$ and $\overline{BA} = \overline{BC}$; therefore, $\overline{AB} = \overline{BC} = \overline{CA}$, proving that the required triangle is indeed equilateral.

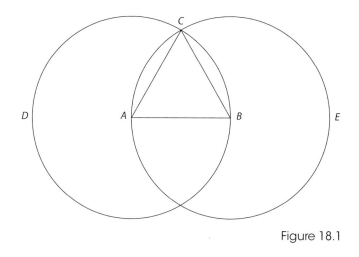

Figure 18.1

We should note that Euclid's compass was different from the familiar modern compass: it "collapsed" when lifted from the paper and, therefore, could not be used to transfer line segments from one place to another. We do not know whether Euclid actually used it as a physical tool or whether he intended it merely as an abstract device with which one could do the construction *in principle*. Whichever the case, constructions with the straightedge and compass—jointly known as the Euclidean tools—

Plate 18. *Seven Circles a Flower Maketh*

became the subject of endless fascination, "a *geometric solitaire* which over the ages has attracted hosts of players, and though now well over two thousand years old, has lost none of its singular charm and appeal."[1]

In fact, you don't even need a straightedge. As the Danish geometer Jørgen Mohr (1640–1697) proved in 1672, every construction that can be done with a straightedge and compass can be done with a compass alone—provided you think of a line as given by the two intersection points of a pair of circles. Because Mohr published his result in Danish rather than Latin—the language of scientific discourse at the time—it received little attention until it was rediscovered in 1797 by the Italian Lorenzo Mascheroni (1750–1800). It was only by a curious incident that Mohr's original theorem came to light, when a young mathematics student found a copy of his work in a secondhand bookstore in Copenhagen.

The result is now known as the Mohr-Mascheroni theorem.

By itself, a geometric construction is a stark, black-and-white array of lines and circles. But add color to it, and it can become an exquisite work or art, as plate 18, *Seven Circles a Flower Maketh*, shows. In the coming chapters we will look at some particular constructions, among them the regular pentagon.

NOTE:

1. The quote is from Howard Eves, *A Survey of Geometry*, p. 154. It might be supposed that the modern compass, being capable of transferring distances, can do more than its collapsible predecessor, but this is not the case: it can be shown that the two are completely equivalent in the sense that each can do everything the other can, although perhaps requiring more steps. For a proof, see Eves, p. 155.

19

Hexagons

A *regular polygon* is a convex polygon whose sides all have the same length and meet each other at the same angle. Next to the equilateral triangle, the simplest regular polygon to construct—using only the Euclidean tools—is the six-sided hexagon. Let the side $\overline{AB}$ be given (figure 19.1). Draw a circle with center at A and radius $\overline{AB}$, place the point of your compass at B, and without changing the compass's opening, swing an arc, cutting the circle at C. Now place your compass at C and, with the opening still the same, swing a second arc, cutting the circle at D. Repeat the process three more times, resulting in points E, F, and G (if you do it one more time, the last point should coincide with B—provided, of course, that your drawing was exact). With a straightedge, connect pairs of adjacent points and you get a perfect hexagon, with its high degree of symmetry (six 60° rotations and six reflections).

The hexagon is one of only three regular polygons that can tile, or *tessellate* the plane—fill it completely without gaps or overlaps. The other two are the square and the equilateral triangle; but since a hexagon can be dissected into six equilateral triangles, the hexagonal and triangular tessellations are not really different. Hexagonal tiling, while not as common as square tiling, can be seen at many places;

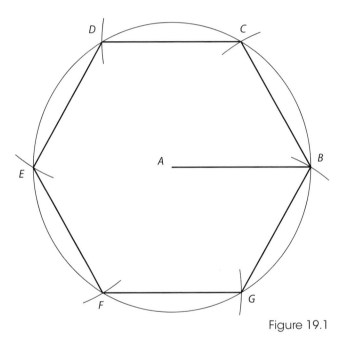

Figure 19.1

a good example is the paving of the subway stations of Washington, DC.

Suppose you want to place rows of identical coins on a table so that each coin touches its immediate neighbors. This can be done in two ways: either the coins in successive rows are placed above one an-

Plate 19. *Parquet*

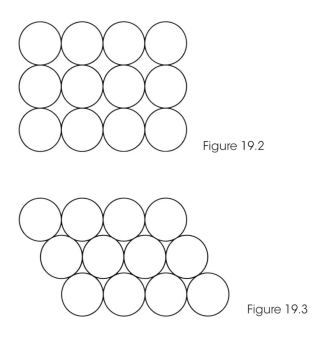

Figure 19.2

Figure 19.3

spheres, resulting in a density of $\pi / \sqrt{18}$ spheres per unit volume, or slightly more than 74 percent. Amazingly, his conjecture—well known to any fruit vendor who packs a pile of oranges in a box—remained unproved until 1998, when the American mathematician Thomas Hales used a computer program to exhaust the large number of possible cases.

The hexagon was known to humans for thousands of years, as evidenced by the six-spiked wheels of Babylonian and Egyptian chariots. Nature, too, takes advantage of the hexagon's high degree of symmetry. Snowflakes, with their infinite variety of fine structure, invariably crystallize into perfect hexagons, a fact that never fails to fascinate nature lovers. And then there is the honeycomb, whose occupants, the hardworking honeybees, diligently shape their wax-made habitat into hexagonal prisms, making this shape a fitting logo on honey products.

Plate 19, *Parquet*, seems at first to show a stack of identical cubes, arranged so that each layer is offset with respect to the one below it, forming the illusion of an infinite, three-dimensional staircase structure. But if you look carefully at the cubes, you will notice that each corner is the center of a regular hexagon. I still remember my amazement when, as a physics student, I first noticed this sixfold symmetry in an object that at first thought should only have two- and fourfold symmetries. It takes a while for the eye to recognize this, as the entire array seems to jump in and out of space, at one moment appearing as if the cubes point straight at you, only to reverse their orientation at the next. An optical illusion? A trick that the brain plays with our eyes? I let the reader decide.

other, so that each coin is surrounded by eight neighbors whose centers form a square (figure 19.2); or else each row is offset with respect to the row below it, the coins in the second row partially filling the space between those of the first row (figure 19.3). In this second arrangement, every coin is surrounded by six neighbors, whose centers form a hexagon. The resulting array is not only more stable but also more efficient—it packs more coins per unit area.

The analogous configuration in *three* dimensions, with spheres replacing the coins, has puzzled mathematicians for more than three hundred years. The great German astronomer Johannes Kepler (1571–1630) conjectured in 1611 that this arrangement represents the most efficient packing of identical

20

Fibonacci Numbers

Almost anyone with the slightest interest in mathematics will be familiar with the name Fibonacci. Leonardo of Pisa—he later adopted the name Fibonacci (son of Bonacci)—was born in Pisa around 1170, the son of a wealthy merchant. Pisa at that time was an important commercial center, serving both Christian Europe and Moslem Middle East and North Africa. Fibonacci was thus acquainted with the newly invented Hindu-Arabic numeration system, with the numerals (or "ciphers") 0 through 9 as its centerpiece. Convinced that this system was superior to the cumbersome Roman numerals, he wrote a book entitled *Liber Abaci* ("Book of the Calculation," sometimes translated as "Book of the Abacus"), in which he advocated the new system and explained its operation. Published in 1202, it became an instant bestseller and was in no small measure responsible for the acceptance of the new system by European merchants and, eventually, by most of the learned world.

So it is ironic that Fibonacci's name is remembered today not for the main thrust of his influential book—promoting the Hindu-Arabic numeration system—but for a little problem he posed in it, perhaps as a recreational exercise. The problem deals with the number of offsprings a hypothetical pair of rabbits can produce, assuming that a pair becomes productive from the second month on and gives birth to a new pair every subsequent month. This leads to the sequence of numbers 1, 1, 2, 3, 5, 8, 13, 21, 34, . . ., in which each number from the third term on is the sum of its two predecessors.[1] The Fibonacci sequence, as it became known, grows very fast: the tenth member is 55, the twentieth is 6,765, and the thirtieth is 832,040. In his famous problem Fibonacci asked how many rabbits will there be after one year. The answer is 144, the twelfth Fibonacci number.

Fibonacci hardly could have anticipated the stir his little puzzle would create. The sequence enjoys numerous properties—so many, in fact, that a scholarly journal, the *Fibonacci Quarterly*, is entirely devoted to it. Fibonacci numbers seem to appear where you least expect them. For example, the seeds of a sunflower are arranged in two systems of spirals, one winding clockwise, the other, counterclockwise. The number of spirals in each system is always a Fibonacci number, typically 34 one way and 55 the other (see figure 20.1), with occasional higher numbers. Smaller Fibonacci numbers also show up in the scales arrangement of pinecones and the leaf patterns of many plants.

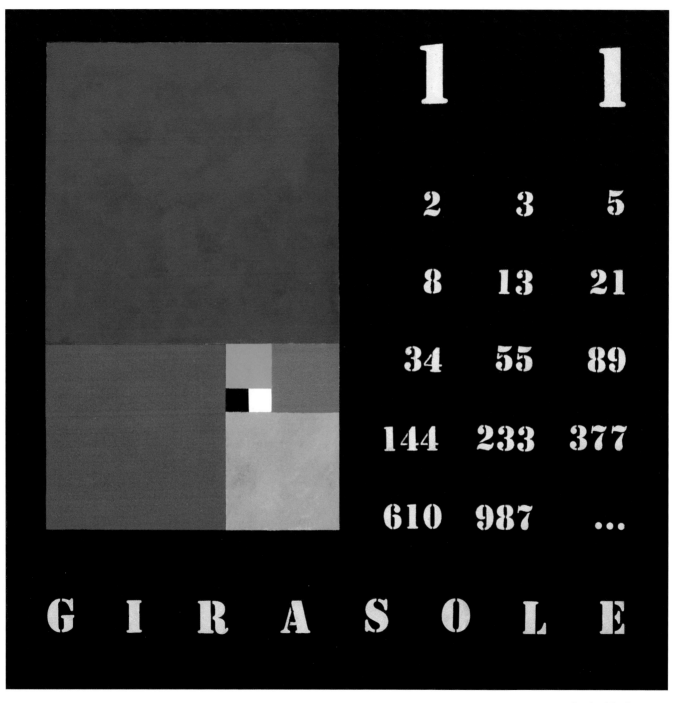

Plate 20. *Girasole*

Figure 20.1

Among the purely mathematical properties of the Fibonacci numbers, we mention here just one: the sum of the first n members of the sequence is always equal to the next-to-next member, minus 1; that is,

$$F_1 + F_2 + F_3 + \cdots + F_n = F_{n+2} - 1.$$

For example, the sum of the first 8 Fibonacci numbers is the tenth number minus 1: $1+1+2+3+5+8+13+21 = 54 = 55-1$. You can use this fact to surprise your friends by asking them to find the sum of, say, the first 10 Fibonacci numbers. Most likely they will start by adding the terms one by one, a process that will take some time. But knowing that the twelfth Fibonacci number is 144, you can outdo them by announcing the answer, 143, while they are still doing their sums. It always works! (See the appendix for a proof.)

Perhaps most surprising of all is a discovery made in 1611 by Johannes Kepler: divide any member of the sequence by its immediate predecessor. As you do this with ever-increasing numbers, the ratios seem to converge to a fixed number, a limit:

$$\frac{2}{1} = 2, \quad \frac{3}{2} = 1.5, \quad \frac{5}{3} = 1.666\ldots, \quad \frac{8}{5} = 1.6,$$

$$\frac{13}{8} = 1.625, \quad \frac{21}{13} = 1.615\ldots, \quad \ldots$$

This limit, about 1.618, turns out to be one of the most famous numbers in mathematics, nearly the equal in status to π and e. Its exact value is $(1+\sqrt{5})/2$. It came to be known as the *golden ratio* (*sectio aura* in Latin), and it holds the secret for constructing the regular pentagon, as we will see in chapter 22.

Plate 20, *Girasole*, shows a series of squares, each of which, when adjoined to its predecessor, forms a rectangle. Starting with a black square of unit length, adjoin to it its white twin, and you get a 2×1 rectangle. Adjoin to it the green square, and you get a 3×2 rectangle. Continuing in this manner, you get rectangles whose dimensions are exactly the Fibonacci numbers. The word *Girasole* ("turning to the sun" in Italian) refers to the presence of these numbers in the spiral arrangement of the seeds of a sunflower—a truly remarkable example of mathematics at work in nature.

NOTE:

1. The sequence is sometimes counted with 0 as the first member. It can also be extended to negative numbers: . . . 5, –3, 2, –1, 1, 0, 1, 1, 2, 3, 5, ...

21

The Golden Ratio

Suppose you are being asked to divide a line segment into two parts such that the whole segment is to the longer part as the longer part is to the shorter. The Greeks were greatly intrigued by this seemingly simple problem, but exactly why is not quite clear: perhaps it was posed by an anonymous scholar as an exercise to his students, or it may have arisen from the challenge of constructing a regular pentagon with straightedge and compass (see the next chapter). Whatever its origins, this particular division of a line segment into two parts became known as the *golden section* (*sectio aura* in Latin). The ratio between the lengths of the two parts is called the *golden ratio* and is usually denoted by the Greek letter φ (phi), although some authors denote it by τ (tau).

Let the line segment be of unit length (figure 21.1). Denoting the length of the longer part by x, the problem leads to the equation

$$\frac{1}{x} = \frac{x}{1-x},$$

which, after rearranging, yields the quadratic equation $x^2 + x - 1 = 0$. This equation has two solutions, one positive and one negative; but since x stands for length, it cannot be negative. Using the familiar quadratic formula and taking only the positive solution, we get

$$x = \frac{-1 + \sqrt{5}}{2},$$

or about 0.618. The golden ratio φ, by definition, is $1/x$. A little arithmetic manipulation will show that $1/x = (1 + \sqrt{5})/2$, which, you will notice, differs from x by exactly 1. Thus, the decimal value of φ is about 1.618.

The number φ enjoys many interesting properties, some quite surprising. We have already noticed that $\varphi = 1 + 1/\varphi$. Multiplying both sides of this equation by φ results in $\varphi^2 = \varphi + 1$. Multiplying again by φ gives us $\varphi^3 = \varphi^2 + \varphi = (\varphi + 1) + \varphi = 2\varphi + 1$. This process can be repeated; replacing φ^2 by $\varphi + 1$ at each step and collecting like terms, we get the sequence:

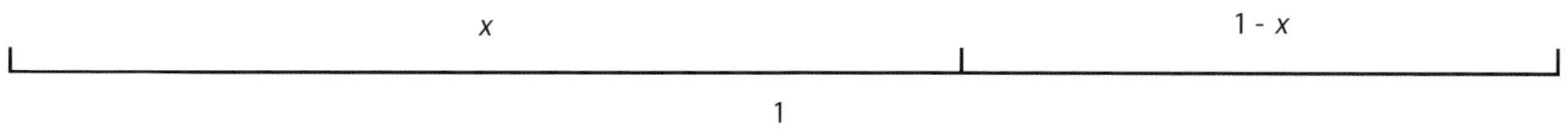

x

$1 - x$

1

Figure 21.1

1, 1, 2, 3, 5, 8, 13, 21, 34, 55, 89, 144, 233, 377, 610, 987, 1597, 2584, 4181, 6765, 10948, 17713, 28661, ...

28661 : 17713 = 1.618...
17713 : 28661 = 0.618...

1.6180339887498948482045868343656381177203091798057628621354486227052604628189024497072072041893911374847540880753868917521266338622235369317931800607667263544333890865959395829056383226613199282902678806752087668925017...

$$\varphi^1 = 1\varphi$$
$$\varphi^2 = 1\varphi + 1$$
$$\varphi^3 = 2\varphi + 1$$
$$\varphi^4 = 3\varphi + 2$$
$$\varphi^5 = 5\varphi + 3$$
$$\varphi^6 = 8\varphi + 5$$
...

$$\varphi = \sqrt{1 + \sqrt{1 + \sqrt{1 + \sqrt{1 + ...}}}}$$

a : b = (a + b) : a
>
a = 61.8 %   b = 38.2 %

$$\varphi = \frac{1 + \sqrt{5}}{2}$$

Plate 21. *The Golden Ratio*

$$\varphi^2 = \varphi + 1, \ \varphi^3 = 2\varphi + 1, \ \varphi^4 = 3\varphi + 2, \ \varphi^5 = 5\varphi + 3,$$
$$\varphi^6 = 8\varphi + 5, \dots.$$

The coefficients in these expressions turn out to be none other than the Fibonacci numbers! In the previous chapter we saw that the ratio of two consecutive Fibonacci numbers approaches the golden ratio as we move higher up in the sequence; now we have a second example of how seemingly unrelated mathematical objects may in fact be intimately connected. Who would have expected that the golden section—a purely geometric entity—would have anything to do with the Fibonacci numbers, whose origin is in number theory?

The golden section has also found its way into art and architecture. It has been claimed that the Greeks, in their obsessive quest for aesthetic perfection, have examined rectangles of various proportions and found that the rectangle whose length-to-width ratio is equal to the golden ratio appeared to be the most aesthetically pleasing. They may have used this proportion in the construction of their temples. The famous Parthenon in Athens has the approximate proportions of the golden ratio, but whether it was built specifically with this ratio in mind is the subject of an ongoing debate.[1] Leonardo da Vinci noticed that if a man stretches his hands to their full extent eagle-manner, the ratio of his height to the wingspan of his hands is approximately 1.6—very close to φ. And in our own time, someone with a keen eye has noticed that the dimensions of a standard credit card are 8.5×5.3 cm, resulting in a length-to-width ratio of 1.60377—within less than 1 percent of φ.

But how close is close? Is 3 a close approximation to π? If so, then we can find a connection to π whenever 3 shows up, which is to say, everywhere. Once

you accept approximations into the picture, you are opening the door to endless speculation, which is indeed what happened with the golden ratio. Regardless of whether its connections to art and architecture are purely accidental or are anchored in some hidden principle, they have greatly added to the aura of mystique that surrounds this number. No wonder medieval scholars dubbed it the *divine proportion*.

Plate 21 showcases a sample of the many occurrences of the golden ratio in art and nature. Some of the panels describe scenes mentioned earlier in this chapter. For the remaining panels, we'll use a "coordinate" notation (x, y) to identify each panel, x standing for the row number (counting from top to bottom) and y for the column (from left to right).

In panel $(1, 1)$ we see a regular pentagon and its associated pentagram, with a smaller pentagon nested inside; as we will see in the next chapter, the golden ratio is the key to constructing a regular pentagon with straightedge and compass.

Panel $(2, 1)$ shows one way of dividing a line segment (here of length 2) in the ratio $\varphi{:}1$; the dot on the base line marks the point of division. Panel $(4, 2)$ shows another way; the horizontal line is parallel to the triangle's base and cuts the two lateral sides at their midpoints.

Panel $(2, 3)$ is rather intriguing. We see an infinite succession of nested square roots, with 1 added to each radical before taking the next, like a Russian *Matryoshka* doll inside which resides a smaller doll, a still smaller one inside that, and so on. Surprisingly, this infinite expression converges to φ.[2]

Panels $(3, 1)$, $(3, 2)$, and $(3, 3)$ bring the golden ratio into three-dimensional space. Take three identical cards, each of length-to-width ratio equal to φ. Cut a slit along the centerline in two of the cards, equal in length to the card's width. Insert the cards into each other as shown, resulting in a three-dimensional structure with 12 corners. Connect pairs of adjacent corners, and you get an icosahedron, a polyhedron with 20 equilateral triangles as faces— one of the five regular solids we met in chapter 4. Interestingly, the two structures appear on the logo of two organizations devoted to mathematical research and teaching—the stacked cards as the icon of the German DFG Research Center Matheon and the icosahedron on the logo of the Mathematical Association of America.

NOTES:

1. On this subject see Livio, *The Golden Ratio*.

2. To see this, let us denote the entire expression by x (assuming, of course, that the nested radicals indeed converge to a limit). Since the same x also appears inside the radical, we have the equation $\sqrt{1+x} = x$. Squaring both sides and rearranging terms, we get the quadratic equation $x^2 - x - 1 = 0$, whose positive solution is $(1+\sqrt{5})/2$, the golden ratio (we take only the positive solution, because by definition the symbol $\sqrt{a}$ is the *positive* solution of the equation $x^2 = a$).

22

The Pentagon

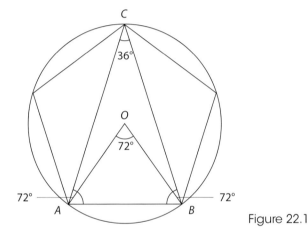

Figure 22.1

The five-sided pentagon has fascinated mathematicians for generations, and it still fascinates them today. In contrast to the three-, four-, and six-sided regular polygons, it is not at all obvious how to construct a regular pentagon, a fact that presented a challenge to the Pythagoreans. The secret to its construction is the 72–72–36-degree triangle *ABC* formed by any of the pentagon's five sides and the vertex opposite to it (figure 22.1).[1] This triangle is known as the *golden triangle* because its side-to-base ratio is exactly the golden ratio $\varphi = (1+\sqrt{5})/2$. Since this ratio can be constructed using a straightedge and compass, the pentagon can be constructed too (we give the proof and construction in the appendix).

The pentagon is related to the golden ratio in other ways as well. For example, if you connect its five vertices by straight lines, you get a *pentagram*, a five-cornered star polygon, inside which resides a smaller pentagon, an exact replica of the original but with side φ^2 times smaller. This process can be repeated, generating ever smaller pentagons and pentagrams whose sides decrease as $1/\varphi^2$, $1/\varphi^4$, $1/\varphi^6, \ldots$ (see plate 22). No wonder the Pythagoreans were enchanted by the pentagram, with its many hidden secrets—so much so that they chose it as their emblem.

In medieval times fortresses were often built in a pentagonal shape, supposedly because it afforded the best firing field from the watchtowers at its corners. But the most famous pentagonal fortress is of modern vintage: the Pentagon in Washington, DC. Theories abound as to why this unusual shape was chosen for the world's largest defense department: economy in building materials, efficient use of space, or a tribute to the ancient Pythagorean emblem. The truth is more prosaic: when construction began in 1941, the only suitable lot within reasonable dis-

Plate 22. *Pentagons and Pentagrams*

tance from the capital was tucked between the Potomac River, one of its tributaries, and three access roads, an area roughly forming a pentagonal shape. So a pentagon it would be, prompting Hermann Weyl, in his classic book *Symmetry*, to comment: "By its size and distinctive shape, it provides an attractive landmark for bombers." Fifty years after he wrote those words, on September 11, 2001, his premonition tragically came true.

The pentagon also made history of a different kind. When the Soviet spacecraft *Luna 2* crashed on the Moon on September 14, 1959, it scattered around a cluster of small pentagonal-shaped metal pennants with the emblem of the Soviet Union, making the pentagon the first human-made object to reach another world.

NOTE:

1. To see why this is a 72–72–36-degree triangle, inscribe the pentagon in a circle with center O. We have $\angle AOB = 360°/5 = 72°$, so $\angle ACB = 36°$ because its vertex is on the circumference and subtends the same arc, $\overset{\frown}{AB}$, as $\angle AOB$.

23

The 17-Sided Regular Polygon

Once we have constructed a regular polygon of n sides—an n-gon, for short—it is easy to construct a regular polygon with twice as many sides—a $2n$-gon: inscribe the n-gon in a circle with center at O (figure 23.1 shows this for the hexagon). Let A and B be two adjacent vertices of this n-gon. Bisect $\angle AOB$ and extend the bisector until it meets the circle at P. Connect P to either A or B, and you have one side of the $2n$-gon. So from the 3-, 4-, and 5-sided gons we can get the 6-, 8-, 10-, and 12-sided gons. Add to that the 15-sided gon (for reasons to be explained shortly), and you have the complete list of regular n-gons the Greeks were able to construct with Euclidean tools.

Imagine the surprise when 19-year-old Carl Friedrich Gauss, just beginning his academic career, announced in 1796 that he could construct a regular gon of 17 sides! Gauss (1777–1855) would soon be recognized as one of the greatest mathematicians of all time, but when he made his sensational discovery he was still unknown. In fact, he had previously planned to become a linguist, but his surprising discovery convinced him that mathematics was his true calling, and so it was.

Before you attempt to construct a 17-sided polygon inscribed in a unit circle, you should be warned

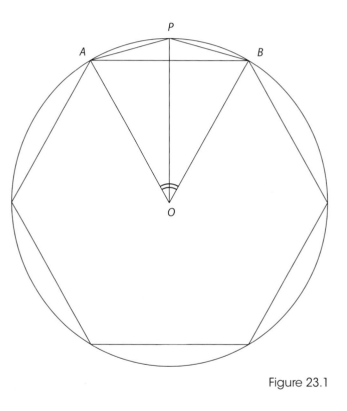

Figure 23.1

of the daunting task awaiting you: the length of each side of your 17-gon is given by the formula

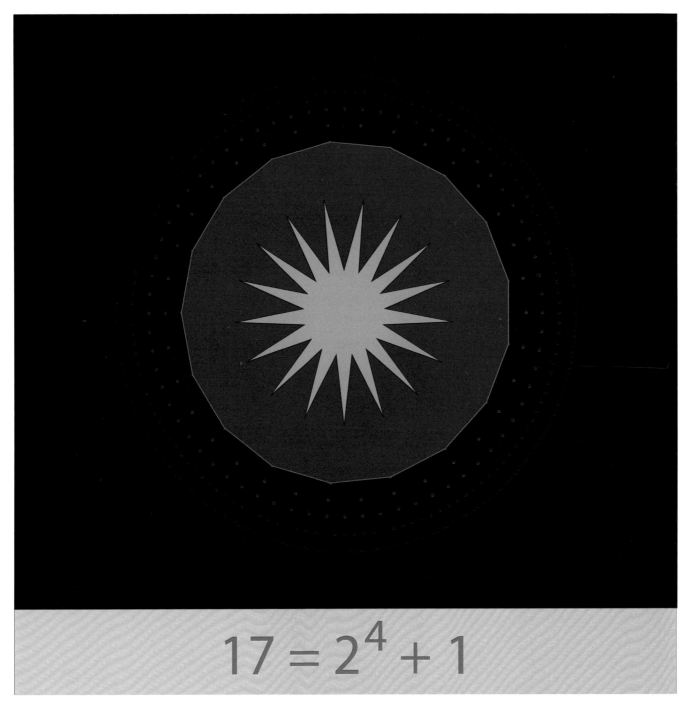

$$17 = 2^4 + 1$$

Plate 23. *Homage to Carl Friedrich Gauss*

$$\frac{1}{4}\sqrt{34 - \sqrt{17} - \sqrt{34 - 2\sqrt{17}} - 2\sqrt{17 + 3\sqrt{17}}}$$
$$\overline{\overline{+ \sqrt{170 - 26\sqrt{17}} - 4\sqrt{34 + 2\sqrt{17}}}} \,,$$

an expression that is sure to give the chills to anyone thinking of pursuing a career in mathematics![1]

Gauss actually did more: he showed that it is possible, in principle, to construct a regular n-gon if n is a product of a nonnegative power of 2 and *distinct* primes of the form $F_k = 2^{2^k} + 1$, where k is a nonnegative integer. Primes of this form are known as *Fermat primes*, named after the great French number theorist Pierre de Fermat (1601–1665). Fermat conjectured that this expression yields a prime for *every* nonnegative k. Indeed, for $k = 0$, 1, 2, 3, and 4 we get $F_k = 3$, 5, 17, 257, and 65,537—all primes. But in 1732 Leonhard Euler proved Fermat wrong: for $k = 5$ we get $2^{2^5} + 1 = 4,294,967,297 = 641 \times 6,700,417$, a composite number. To this day it is not known if any other Fermat primes exist, so it is possible that there are other, as yet undiscovered, regular gons that can be constructed with the Euclidean tools. Needless to say, if such polygons do exist, they must have a huge number of sides, making any *practical* construction totally out of the question. Gauss, however, was not so much concerned about the actual construction; what mattered to him was the laws that govern such a construction.

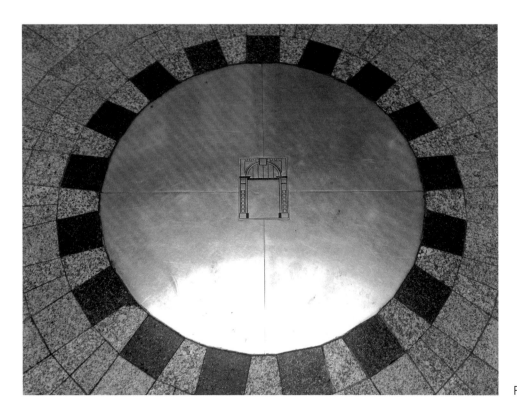

Figure 23.2

Gauss's achievement is immortalized in his German hometown of Brunswick, where a large statue of him is decorated with an ornamental 17-pointed star (plate 23 is an artistic rendition of the actual star on the pedestal, which has deteriorated over the years); reportedly the mason in charge of the job thought that a 17-sided *polygon* would look too much like a circle, so he opted for the star instead.[2] In 1837, Pierre Laurent Wantzel (1814–1848) proved that the polygons Gauss identified are in fact the *only* ones constructible with Euclidean tools, making Gauss's discovery a necessary and sufficient condition for the construction. Thus, a regular polygon of 15 sides is constructible, because $15 = 3 \times 5$, and both 3 and 5 are Fermat primes $(3 = 2^{2^0} + 1$ and $5 = 2^{2^1} + 1);$[3] but a regular 14-sided gon is not, because $14 = 2 \times 7$, and 7 is not a Fermat prime. Neither is a 50-sided gon, because $50 = 2 \times 5 \times 5$, and the double appearance of 5 in the factorization disqualifies the 50-gon from being constructible.

It is extremely rare to find examples of a 17-sided regular gon in nature, let alone in art or architecture. But if you search around long enough, you will eventually find what you were looking for. One of us (Jost) recently discovered in a shopping mall in the town of Leipzig a 17-sided glass dome, with a 17-sided pattern decorating the floor under it (figure 23.2).

NOTES:

1. For further discussion of the 17-gon, see Hartshorne, *Geometry*, chapter 29.

2. I wish to thank Professor Manfred Stern of Halle, Germany, for updating me about the Gauss statue in Brunswick.

3. Proposition 16 of Book IV of the *Elements* gives the construction of a regular 15-sided polygon. For a nice animation of the construction, see http://en.wikipedia.org/wiki/Pentadecagon#Regular_pentadecagon.

$$24$$

Fifty

Number aficionados will tell you that every number has its own personality, its special features and its unique meaning—although people might differ as to what exactly that meaning is. But from the perspective of a number theorist, what matters most is a number's prime factors. This prime factorization is the key to most, if not all, the mathematical properties of a number.

Let us take a look at 50—a nice, round number, the halfway point of a century, the age at which many of us start to think about our mortality. As decreed in the Bible (Leviticus 25:10), the year following a 49-year-long cycle, the year of the *Jubilee* (from the Hebrew *Yovel*) was designated as a time of renewal, of emancipation of slaves and restoration of leased lands to their former owners. To this day a fiftieth anniversary is known as a jubilee.

The prime factorization of 50 is $2\times5\times5$, and as we saw in the previous chapter, the double presence of the Fermat prime 5 in the factorization rules out the possibility of constructing a regular polygon of 50 sides with Euclidean tools. Interestingly, a 51-gon—practically indistinguishable from its 50-sided neighbor—*is* constructible with Euclidean tools; that's because $51=3\times17$, and both 3 and 17 are Fermat primes $(3 = 2^{2^0} +1$ and $17 = 2^{2^2} +1)$.

Fifty is also the smallest number that can be written as the sum of two squares *in two different ways*: $50=5^2+5^2=7^2+1^2$. In his book *Arithmetica*, Diophantus of Alexandria (probably the third century CE) showed that the product of two numbers, each of which is a sum of two squares, is again a sum of two squares:

$$(a^2 + b^2)(c^2 + d^2) = a^2c^2 + a^2d^2 + b^2c^2 + b^2d^2$$
$$= (a^2c^2 + 2abcd + b^2d^2)$$
$$+(a^2d^2 - 2abcd + b^2c^2)$$
$$= (ac + bd)^2 + (ad - bc)^2.$$

But the last expression can also be written as $(ac-bd)^2+(ad+bc)^2$, showing that the said product can be expressed as the sum of two squares in two different ways.[1] Since 50 is the product of 5 and 10, each of which is a sum of two squares $(5=1^2+2^2$ and $10=1^2+3^2)$, we have

$$50=(1\times1+2\times3)^2+(1\times3-2\times1)^2=7^2+1^2,$$

and, again,

$$50=(1\times1-2\times3)^2+(1\times3+2\times1)^2=5^2+5^2.$$

The next such number is $65=5\times13=8^2+1^2=7^2+4^2$.

Plate 24.1. *Stars and Stripes*

Plate 24.2. *Celtic Motif 2*

Coincidentally, 50 stars appear on the American national flag, the *Stars and Stripes*, each star representing one state of the Union. This opens up an interesting possibility. Most people, looking at the pattern of stars on the flag, will see an arrangement of nine horizontal rows alternating between six and five stars per row ($5 \times 6 + 4 \times 5 = 50$). But if you look at the star arrangement diagonally, an entirely new pattern emerges: five rows with 1, 3, 5, 7, and 9 stars, followed by the same pattern in reverse. In chapter 4 we saw that the sum of the first n consecutive odd integers is n^2. In our case, $1 + 3 + 5 + 7 + 9 = 5^2 = 25$, making the total number of stars 50.

But 50 is also equal to $7^2 + 1^2$, which can be arranged in a square of $7 \times 7 = 49$ stars and a single additional star anywhere outside. This single star could then stand for any one state of the Union, allowing each state to claim that it has a privileged status

without offending any other state! We display the two flags, the actual and the hypothetical, in plate 24.1.

Plate 24.2 shows a laced pattern of 50 dots, based on an ancient Celtic motif. Note that the entire array can be crisscrossed with a single interlacing thread; compare this with the similar pattern of 11 dots (see chapter 17), where two separate threads were necessary to cover the entire array. As we said before, every number has its own personality.

NOTE:

1. Except when $a = b$ or $c = d$, in which case the product is the sum of two squares in just one way; for example, $10 = 2 \times 5 = (1^2 + 1^2)(1^2 + 2^2) = (1 \times 1 \pm 1 \times 2)^2 + (1 \times 2 \mp 1 \times 1)^2 = 3^2 + 1^2$.

25

Doubling the Cube

According to legend, at one time the Greek town of Delos was afflicted by a devastating plague that nearly decimated its population. In desperation, the city elders consulted the oracles, who determined that the god Apollo was unhappy with the small size of the pedestal on which his statue was standing. To appease him, they recommended to double the volume of the cubical pedestal. The task was given to the town's mathematicians, who soon realized that doubling the *side* of the cube would not do it—it would increase the volume *eightfold* and would make the pedestal unreasonably large. What to do?

It took mathematicians some two thousand years to realize that the problem could not be solved with Euclidean tools, no matter how much one tried. That is, given a line segment of length a, no construction, using only straightedge and compass, could produce a line segment of length x such that $x^3 = 2a^3$. To be sure, a variety of other tools have been devised to accomplish the task, but they were regarded as "mechanical" and thus not befitting a "true" solution in accordance with Plato's decree. (Of course, the straightedge and compass are mechanical tools too, but for tradition's sake they are the only tools permitted in the construction.)

The *Delian problem*, as it has been known since, was not the only problem that couldn't be solved with Euclidean tools. Two other constructions caused mathematicians even greater headache: trisecting an arbitrary angle, and squaring the circle—that is, constructing a square equal in area to that of a given circle. All three problems, collectively known as the *classical problems of antiquity*, were settled only in the nineteenth century—the first two by Pierre Laurent Wantzel in 1837, and the last—squaring the circle—by Ferdinand Lindemann in 1882 (see chapter 26). Using algebraic methods that were not available to the Greeks, these mathematicians set the conditions under which a geometric construction, suitably translated into a set of equations, could be achieved with Euclidean tools. Among the constructions that *cannot* be done is that of a line segment whose length is equal to the cube root of another, given line segment (except, of course, when the latter is a perfect cube); and since the Delian problem leads to the equation $x = \sqrt[3]{2}\ a$, the construction cannot be achieved with Euclidean tools.

The impossibility of solving the three classical problems was not the first time that a mathematical task has been proved impossible to achieve under given restrictive conditions. We recall the intellec-

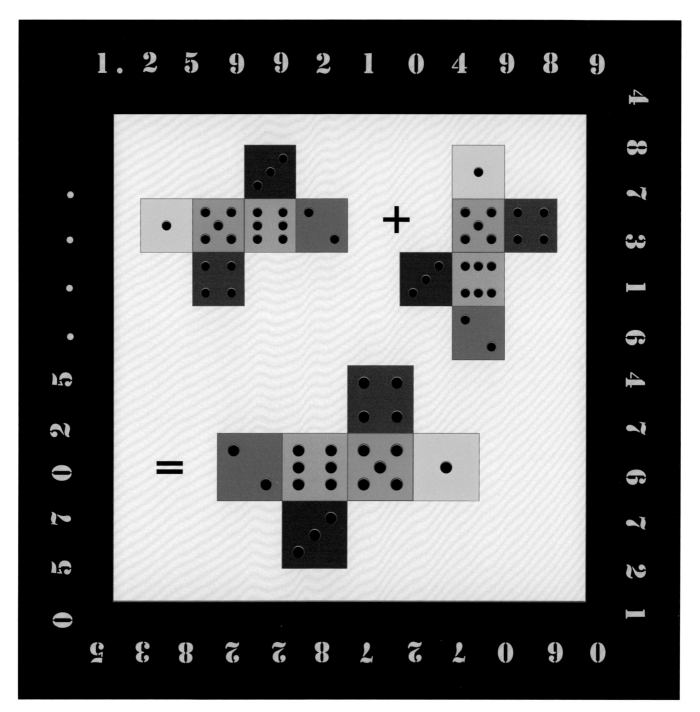

Plate 25. *The Oracles of Delos*

tual crisis that struck the Pythagoreans when they proved that the square root of 2 could not be expressed as a ratio of two integers. it showed that even mathematics, for all its reputation as the discipline of absolute, infallible truth, has its inherent limitations. But this hasn't stopped countless amateurs and cranks from submitting their "solutions" to professional journals, with visions of lasting fame and perhaps even a monetary award. Apparently some people will never take no for an answer—even if that no derives its authority from mathematics!

Plate 25, *The Oracles of Delos*, shows a fanciful "duplication" of two unfolded dice, which, when added together, produce a third dice of twice the volume of the original dice. Of course, the plus and equal signs should not be taken literally, reminding us again that the artist is not bound by the same constraints that limit a mathematician. And yet not everything is fanciful here: the side of the large cube is about $1.259\ldots$ as large as that of the small cubes. This number is the decimal value of $\sqrt[3]{2}$, the very stumbling block that stood in the way of solving the Delian problem.

26

Squaring the Circle

At first glance, the circle may seem to be the simplest of all geometric shapes and the easiest to draw: take a string, hold down one end on a sheet of paper, tie a pencil to the other end, and swing it around—a simplified version of the compass. But first impressions can be misleading: the circle has proved to be one of the most intriguing shapes in all of geometry, if not the most intriguing of them all.

How do you find the area of a circle, when its radius is given? You instantly think of the formula $A = \pi r^2$. But what exactly is that mysterious symbol π? We learn in school that it is approximately 3.14, but its *exact* value calls for an endless string of digits that never repeat in the same order. So it is impossible to find the exact area of a circle numerically. But perhaps we can do the next best thing—construct, using only straightedge and compass, a square equal in area to that of a circle?

This problem became known as *squaring the circle*—or simply the *quadrature* problem—and its solution eluded mathematicians for well over two thousand years. The ancient Egyptians came pretty close: In the Rhind Papyrus, a collection of 84 mathematical problems dating back to around 1800 BCE, there is a statement that the area of a circle is equal to the area of a square of side $\frac{8}{9}$ of the circle's diameter. Taking the diameter to be 1 and equating the circle's area to that of the square, we get $\pi(\frac{1}{2})^2 = (\frac{8}{9})^2$, from which we derive a value of π equal to $\frac{256}{81} \approx 3.16049$—within 0.6 percent of the true value. However, as remarkable as this achievement is, it was based on "eyeballing," not on an exact geometric construction.

In the Bible (I Kings 7:23) we find the following verse: "And he made a molten sea, ten cubits from one brim to the other; it was round all about . . . and a line of thirty cubits did compass it round about." In this case, "he" refers to King Solomon, and the "molten sea" was a pond that adorned the entrance to the Holy Temple in Jerusalem. Taken literally, this would imply that $\pi = 3$, and the quadrature of the circle would have become a simple task! A great deal of commentary has been written on this one verse (it also appears, with a slight change, in II Chronicles 4:2), but that would take us outside the realm of geometry. Plate 26.1, *$\pi = 3$*, quotes this famous verse in its original Hebrew; to read it, start at the red dot and proceed counterclockwise all the way around.

Numerous attempts have been made over the centuries to solve the quadrature problem. Many careers were spent on this task—all in vain. The defini-

Plate 26.1. $\pi = 3$

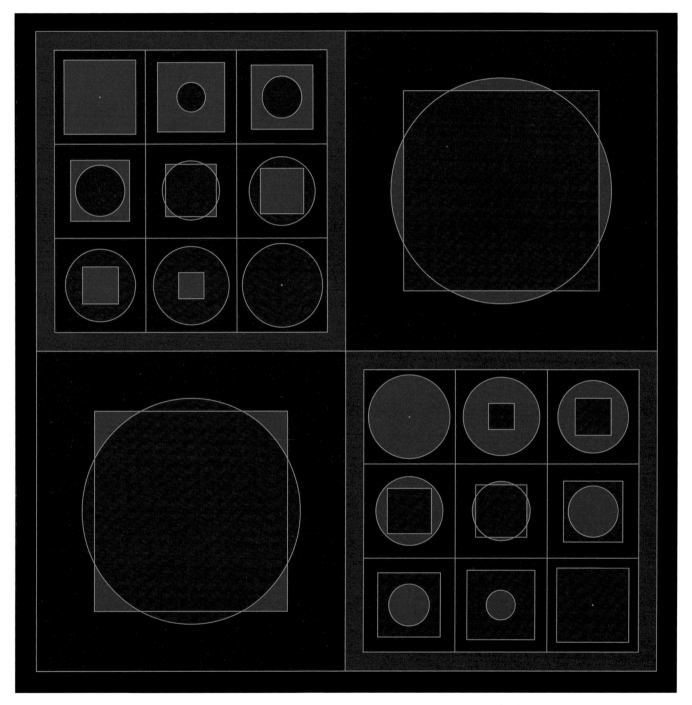

Plate 26.2. *Metamorphosis of a Circle*

tive solution—a negative one—came only in 1882, when Carl Louis Ferdinand von Lindemann (1852–1939) proved that the task cannot be done—it is impossible to square a circle with Euclidean tools. Actually, Lindemann proved something different: that the number π, the constant at the heart of the quadrature problem, is transcendental. A *transcendental number* is a number that is not the solution of a polynomial equation with integer coefficients. A number that is not transcendental is called *algebraic*. All rational numbers are algebraic; for example, $\frac{3}{5}$ is the solution of the equation $5x - 3 = 0$. So are all square roots, cubic roots, and so on; for example, $\sqrt{2}$ is the positive solution of $x^2 - 2 = 0$, and $\sqrt[3]{2 - \sqrt{5}}$ is one solution of $x^6 - 4x^3 - 1 = 0$. The name *transcendental* has nothing mysterious about it; it simply implies that such numbers transcend the realm of algebraic (polynomial) equations.

Now it had already been known that if π turned out to be transcendental, this would at once establish that the quadrature problem cannot be solved. Lindemann's proof of the transcendence of π therefore settled the issue once and for all. But settling the issue is not the same as putting it to rest; being the most famous of the three classical problems, we can rest assured that the "circle squarers" will pursue their pipe dream with unabated zeal, ensuring that the subject will be kept alive forever.

Plate 26.2, *Metamorphosis of a Circle*, shows four large panels. The panel on the upper left contains nine smaller frames, each with a square (in blue) and a circular disk (in red) centered on it. As the squares decrease in size, the circles expand, yet the sum of their areas remains constant. In the central frame, the square and circle have the same area, thus offering a computer-generated "solution" to the quadrature problem. In the panel on the lower right, the squares and circles reverse their roles, but the sum of their areas is still constant. The entire sequence is thus a metamorphosis from square to circle and back.

Of course, Euclid would not have approved of such a solution to the quadrature problem, because it does not employ the Euclidean tools—a straightedge and compass. It does, instead, employ a tool of far greater power—the computer. But this power comes at a price: the circles, being generated pixel by pixel like a pointillist painting, are in reality not true circles, only simulations of circles.[1] As the old saying goes, "there's no free lunch"—not even in geometry.

NOTE:

1. The very first of the 23 definitions that open Euclid's *Elements* defines a point as "that which has no part." And since all objects of classical geometry—lines, circles, and so on—are made of points, they rest on the subtle assumption that Euclidean space is continuous. This, of course, is not the case with computer space, where Euclid's dimensionless point is replaced by a pixel—small, yet of finite size—and space between adjacent pixels is empty, containing no points.

27

Archimedes Measures the Circle

By any measure, Archimedes (ca. 285–212 BCE) is considered the greatest scientist of antiquity, the equal of Newton and Einstein. As with most of the ancient Greek sages, much of what we know about him was written by later historians, who often confused legend with fact; thus many of the stories about him must be taken with a grain of salt. He was born in the town of Syracuse, on the southeast coast of the island of Sicily, where he spent all his life. In popular accounts Archimedes is best remembered for his spectacular engineering feats, but he considered himself first and foremost a pure mathematician, interested in mathematical theorems for their own sake, with little regard for their practical applications. Yet when the Roman navy lay siege to his town, King Heron called upon him to design war machines with which the city could be defended. Among his devices was an enormous crane that could pluck a ship out of the sea, hoist it high in the air, and then let it fall to its doom; he also devised huge concave mirrors that aimed the sun's rays at the Roman fleet and set its ships ablaze. Whether any of these machines was actually built is not known, but their stories made his name legendary.

By far the most famous story about Archimedes is his investigation of the king's crown. Suspicious that the crown was made of base metals rather than pure gold, Heron asked Archimedes to investigate. Immersing the crown in Syracuse's public bath and comparing its weight to that of the spilled water, he concluded that the crown was indeed a forgery. Beside himself with excitement, Archimedes leaped out of the bath and ran down the streets of Syracuse, shouting *eureka*—I found it!

In a rare case of exact dating of an ancient event, Archimedes's year of death is known because it happened when the Romans finally breached Syracuse's walls and took the city by surprise. Their commander, General Marcellus, ordered his troops to capture the renowned scientist alive and treat him with dignity. A soldier came across an old man crouching over some geometric figures drawn in the sand. Being ordered to stand up, the man ignored the soldier, who then drew his sword and killed him: it was Archimedes. The year was 212 BCE.

Archimedes wrote on a wide range of subjects, but only a dozen or so of his works survived. Among them is a small tract with the title *Measurement of a Circle*, in which he devised a method for approximating the value of π to any desired accuracy. His idea was to inscribe regular polygons of 6, 12, 24, 48 and 96 sides inside a circle, find the perimeter of each

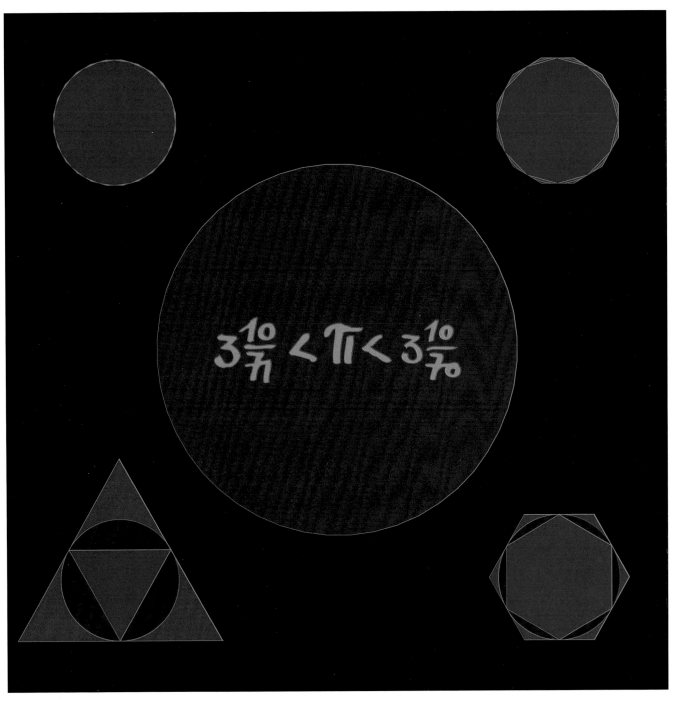

Plate 27. *Homage to Archimedes*

polygon, and divide it by the circle's diameter (by definition, π is the circumference-to-diameter ratio). With each step, the polygons will grip the circle more tightly from within, resulting in approximations of π progressively increasing in accuracy. These approximations, however, are all undervalues of the exact value of π. Archimedes therefore repeated the process with *circumscribing* polygons, gripping the circle from the outside and giving a series of overvalues of increasing accuracy. From the 96-sided inscribed and circumscribing polygons, he concluded that the actual value of π lies between $3\frac{10}{71}$ and $3\frac{10}{70}$ (in decimal notation, between 3.14085 and 3.14286). This last value is equal to $\frac{22}{7}$, an approximation which in precomputer times was often used as a rough estimate.

Besides devising the first workable algorithm for approximating π, Archimedes's method also gave us a glimpse into the theoretical nature of this famous number. It hinted to the fact that the exact value of π can never be found, because it involves a process that must be repeated infinitely many times. It

would be another two thousand years before mathematicians would prove this fact conclusively.

Plate 27, *Homage to Archimedes*, shows a black circle and a series of inscribed and circumscribing regular polygons (in blue and red, respectively) of 3, 6, 12, 24, and 48 sides. We see how the circle is squeezed between each pair of polygons, the fit getting tighter as the number of sides increases. The final (central) circle is practically indistinguishable from the 48-sided polygons that hold it tight. For practical reasons Archimedes started with a hexagon rather than a triangle, because its perimeter is easy to find; and he doubled the number of sides so that he could use a formula he himself had devised for computing the perimeter of a regular $2n$-gon from that of a regular n-gon.[1]

NOTE:

1. For a full account of Archimedes's method, see *The Works of Archimedes*, edited by T. L. Heath (New York: Dover. 1953), pp. 91–98.

28

The Digit Hunters

In the second century BCE, during Archimedes's lifetime, the Hindu-Arabic numeration system was still more than a thousand years in the future. So Archimedes had to do all his calculations in a strange hybrid of the Babylonian sexagesimal (base 60) system and the Greek system, in which each letter of the alphabet had a numerical value (alpha = 1, beta = 2, and so on). Today, of course, we associate the value of π with its decimal expansion—a nonrepeating, seemingly random string of digits that goes on forever. Terminate this expansion after any number of digits, and you'll get only an approximation of π.

So, where should we stop? For many daily practical tasks, the simple fraction $^{22}/_7 \approx 3.1428571$ will suffice, differing from π by just 0.04 percent. The Chinese mathematician Zu Chongzhi (Tsu Ch'ung-Chih, 429–501) around 480 CE discovered the nice approximation $^{355}/_{113} \approx 3.1415929$, accurate to six places—that is, to the nearest millionth. The Dutch-German mathematician Ludolph van Ceulen (1540–1610) computed π to 20 decimal places, using Archimedes's method with polygons of 60×2^{29} sides and spending much of his professional life on the task (he later improved his calculations to 35 places). After Van Ceulen's death, his widow re-

portedly had the number inscribed on his tombstone in Leiden, but all traces of it have been lost. Here it is:

3.14159265358979323846264338327950288.

Apparently nothing would stop the digit hunters from attempting to find π to ever more decimal places. In 2009 a Japanese team set the record at 2.6 *trillion* digits, but it is only a question of time before this record, too, will be broken.[1] Though of no apparent practical use, such enormous strings of digits may shed light on some as yet unanswered questions about π, among them whether it is a *normal* number (a number whose digits follow a uniform distribution; that is, all blocks of digits of given length appear with equal probability, regardless of the base in which the number is expressed).

Not far behind the digit hunters are the digit reciters. In 2006 the Japanese Akira Haraguchi, then 60 years old, set a record by reciting the first 100,000 digits of π from memory, taking him more than 16 hours to read them out. Some feat!

How do the digit hunters achieve their task? There are numerous infinite series and products that allow us to approximate π to any desired accuracy, some doing the job better than others. The first to

3. 14159265358979323846264338327950288419716939937510582097494459230781640628620899862803482534211706798214808651328230664709384460955058223172535940812848111745028410270193852110555964462294895493038196442881097566593344612847564823378678316527120190914564856692346034861045432664821339360726024914127372458700660631558817488152092096282925409171536436789259036001133053054882046652138414695194151160943305727036575959195092861173819326117931051185480744623799627495673518857527248912279381830119491298336733624406566430860021394946395224737190702179860943702770539217176293176752384674481846766940513200056812714526356082778577134275778960917363717872146844090122495343014654958537105079227968925892354201995611212902196086403441815981362977477130996051870721134999999837297804995105973173281609631859502445945534690830264252230825334468503526193118817101000313783875288658753320838142061717766914730359825349042875546873115956286388235378759375195778185778053217122680661300192787661119590921642019893809525720106548586327886593615338182796823030195203530185296899577362259941389124972177528347913151557485724245415069595082953311686172785588907509838175463746493931925506040092770167113900984882401285836160356370766010471018194295559619894676783744944825537977472684710404753464620804668425906949129331367702898915210475216205696602405803815019351125338243003558764024749647326391419927260426992279678235478816360093417216412199245863150302861829745557067498385054945885869269956909272107975093029553211653449872027559602364806654991198818347977535663698074265425278625518184175746728909777727938000816470600161452491921732172147723501414419735685481613611573525521334757418494684385233239073941433345477624168625189835694855620992192221842725502542568876717904946016534668049886272327917860857843838279679766814541009538837863609506800642251252051173929848960841284886269456042419652850222106611863067442786220391949450471237137869609563643719172874677646575739624138908658326459958133904780275900994657640789512694683983525957098258226205224894077267194782684826014769909026401363944374553050682034962524517493996514314298091906592509372216964615157098583874105978859597729754989301617539284681382686838689427741559918559252459539594310499725246808459872736446958486538367362226260991246080512438843904512441365497627807977156914359977001296160894416948685558484063534220722258284886481584560285060168427394522674676788952521385225499546667278239864565961163548862305774564980355936345681743241125150760694794510965960940252288797108931456691368672287489405601015033086179286809208747609178249385890097149096759852613655497818931297848216829989487226588048575640142704775551323796414515237462343645428584447952658678210511413547357395231134271661021359695362314429524849371871101457654035902799344037420073105785390862198848757467919186993935670130918827615367270392926115653488943179...

Plate 28. *Almost* π

devise such a formula was the Frenchman François Viète (1540–1603), who in 1596 discovered the infinite product

$$\frac{2}{\pi} = \frac{\sqrt{2}}{2} \cdot \frac{\sqrt{2+\sqrt{2}}}{2} \cdot \frac{\sqrt{2+\sqrt{2+\sqrt{2}}}}{2} \cdots.$$

In 1671, the Scotsman James Gregory (1638–1675) discovered the infinite series

$$\frac{\pi}{4} = 1 - \frac{1}{3} + \frac{1}{5} - \frac{1}{7} + - \cdots.$$

Not to be outdone, Leonhard Euler (1707–1783) solved one of the great mysteries of his time: to find the sum of the series $1 + \frac{1}{2^2} + \frac{1}{3^2} + \frac{1}{4^2} + \cdots$. In 1734 he announced that the series converges to $\frac{\pi^2}{6}$:

$$\frac{\pi^2}{6} = 1 + \frac{1}{2^2} + \frac{1}{3^2} + \frac{1}{4^2} + \cdots.$$

We listed these formulas here mainly because of their historical significance, although their slow rate of convergence makes them of little use in practice (with the last series, it takes 600 terms to get π to just two decimal places, 3.14). And yet they are remarkable because they tie the number π, the circumference-to-diameter ratio of a circle, to the integers; they show once again the universality of mathematics—its ability to link together different concepts that at first sight seem totally unrelated.

Plate 28, *Almost π*, gives the first 12,827 digits of this famous number, enough to give any potential digit reciter a full load of numbers to memorize.

NOTE:

1. Source: BBC News, on the Web at http://news.bbc.co.uk/2/hi/technology/8442255.stm.

29

Conics

Imagine slicing a cone—for illustration's sake think of it as an ice cream cone—with a swift stroke of a knife. If you slice it in a plane parallel to the cone's base, you get a circular cross section. Tilt the angle slightly, and you get an ellipse. Tilt the angle even more, and the ellipse becomes narrower, until it no longer closes on itself: it becomes a parabola. This happens when the cut is parallel to the side of the cone. Increase the angle yet again, and you get two disconnected curves, the two branches of a hyperbola (provided you regard the cone as a double cone joined at its apex). Taken together, these curves—to which we may add a pair of straight lines when the cut is along the cone's axis—comprise the five *conic sections* (figure 29.1).

The Greek mathematician Apollonius of Perga (ca. 262–190 BCE) wrote an extensive treatise on the conic sections. He gave them the names *ellipse*, *parabola*, and *hyperbola*, according to whether the cut is at an angle smaller than, equal to, or greater than the angle between the cone's base and its side.

The conics are endowed with numerous properties, some shared by the entire family, while others are unique to each member. Let us begin with the parabola. On a sheet of paper draw a line *d* and choose a point *F* not on *d*. The parabola is the set

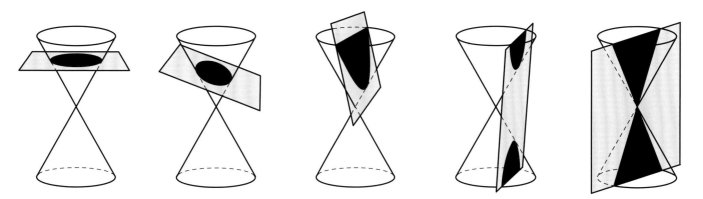

Figure 29.1

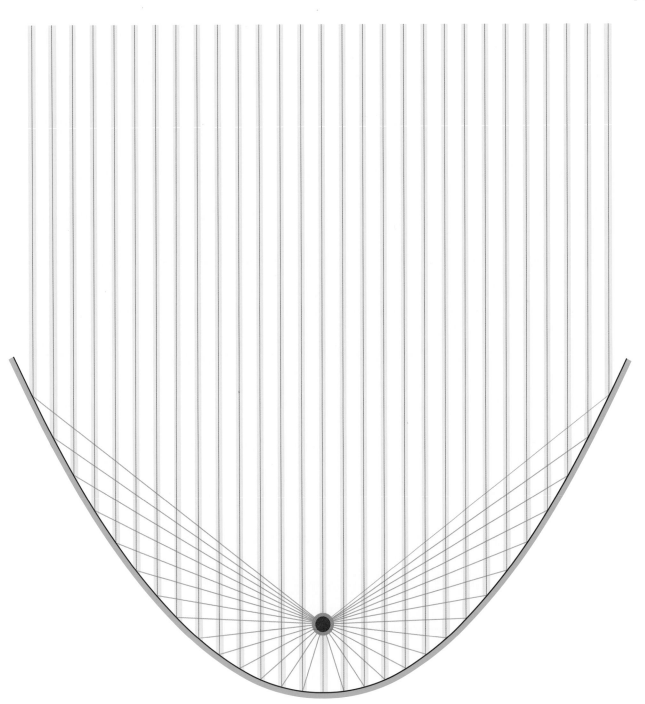

Plate 29.1. *Reflecting Parabola*

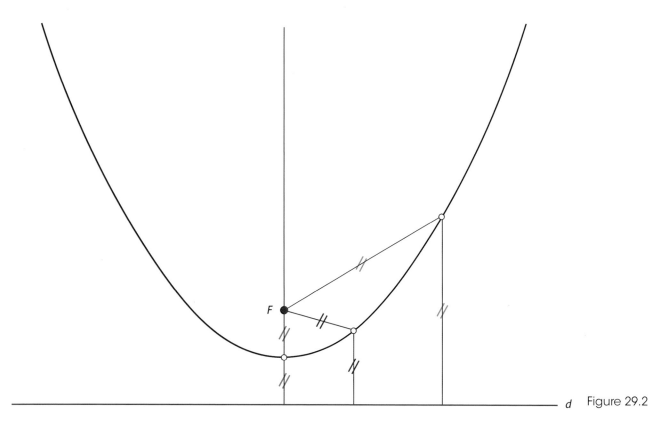

F

d   Figure 29.2

("locus") of all points equidistant from *d* and *F* (figure 29.2). *d* is called the *directrix*, and *F*, the *focus* of the parabola. Like all conics, the parabola is a symmetric curve: its axis of symmetry passes through *F* and runs at a right angle to *d*. As Archimedes had discovered, rays of sunlight arriving at the parabola in a direction parallel to its axis are reflected toward a single point, the focus, as shown in plate 29.1 (indeed, "focus" in Latin means fireplace). This property finds its modern use in the ubiquitous satellite antenna dish—a concave surface with a parabolic cross section that, when aimed at a satellite in geostationary orbit, collects its signals at the focal point, where they are amplified and fed into your TV screen. It also works in reverse: if a source of light or a radio transmitter is placed at the focus, the emitted beam will be reflected in a direction parallel to the parabola's axis, a feature brought to good use in the reflective surface of a car's headlights.

• ◆ •

The ellipse is the locus of points, the sum of whose distances from two fixed points, the ellipse's *foci*, is constant. A ray of light emanating from one focus is reflected by the ellipse toward the other focus. A striking example of this is the famous Whispering Gallery in the United States Capitol in Washington,

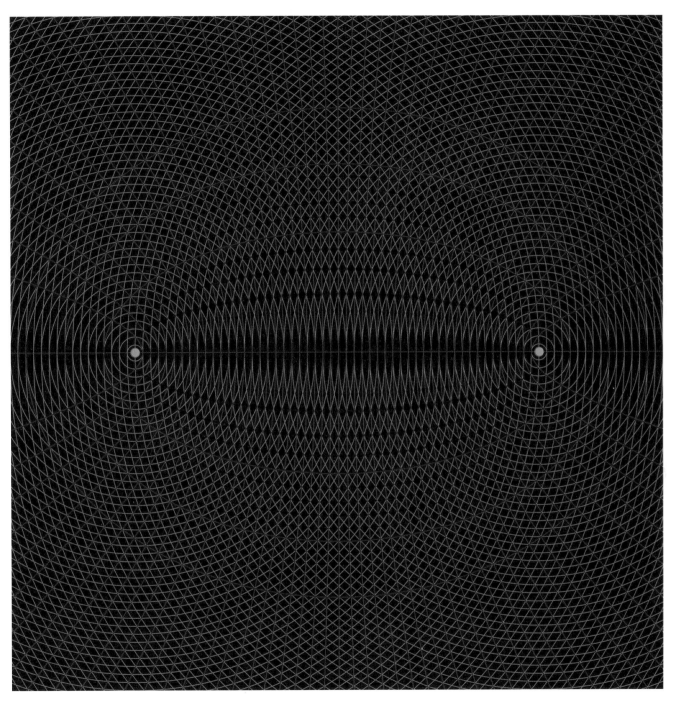

Plate 29.2. *Ellipses and Hyperbolas*

DC, where visitors gathered at one focal point of the elliptically shaped hall can hear their guide, standing some distance away at the other focus, whisper some words seemingly to himself or herself. It never fails to impress the audience.

•◆•

Like the ellipse, the hyperbola has two foci, but this time the *difference* of the distances from any point on the hyperbola to the foci is constant. The hyperbola consists of two disconnected branches, separated by two lines of demarcation that cross each other midway between the foci. These lines are the hyperbola's *asymptotes*: they approach the hyperbola ever so close but never touch it—which is precisely what the word asymptote means: "don't touch me." The asymptotes are like signposts pointing the way to infinity.

A ray of light originating at one focus is reflected by the hyperbola in a direction *away* from the other focus. So the word *focus* in this case is really a misnomer: rather than concentrate at the focus, beams of light are scattered in every direction as they bounce off the hyperbola; "antifocus" would perhaps be a better name.

When you throw two stones into a pond, each will create a disturbance that propagates outward from the point of impact in concentric circles. The two systems of circular waves eventually cross each other and form a pattern of ripples, alternating between crests and troughs. Because this *interference pattern* depends on the phase difference between the two oncoming waves, the ripples invariably form a system of confocal ellipses and hyperbolas, all sharing the same two foci. In this system, no two ellipses ever cross one another, nor do two hyperbolas, but every ellipse crosses every hyperbola at right angles. The two families form an *orthogonal system* of curves, as we see in plate 29.2.

But the conic sections also play a role on a much grander scale. The German astronomer Johannes Kepler, whom we met already twice before, discovered that all planets move around the Sun in elliptical orbits. This discovery finally put to rest the old Greek belief that the planets move around the Earth in perfect circles or combinations of circles. A century later Isaac Newton would show that *every* celestial body—whether a planet, a comet, or a moon—moves around its parent body in an ellipse, a parabola, or a hyperbola. The conic sections thus became the *cosmic sections*.

$$\frac{3}{3} = \frac{4}{4}$$

We encountered the geometric progression in chapter 16 in connection with the runner's paradox. Many interesting results can be obtained using a geometric progression, some quite unexpected. Consider a square of unit side, divide it into four equal smaller squares, and shade the upper-right square, as in figure 30.1. Now divide each of the remaining, unshaded squares into four equal parts, and shade the upper-right quarter of each (figure 30.2). Repeating the process again and again, will the shaded area approach a limit? If so, what is it?

In the first stage there is one shaded square of side ½, so its area is $(½)^2 = ¼$. In the second stage there are three shaded squares, each of side ¼ and area $(¼)^2 = 1/16$. In the third stage there are nine shaded squares, each of area $(⅛)^2 = 1/64$. Continuing in this way, the total shaded area will be

$$\frac{1}{4} + \frac{3}{16} + \frac{9}{64} + \frac{27}{256} + \cdots = \frac{1}{4}\left[1 + \frac{3}{4} + \left(\frac{3}{4}\right)^2 + \left(\frac{3}{4}\right)^3 + \cdots\right].$$

The expression inside the brackets is a geometric series with an initial term 1 and common ratio ¾. Since this common ratio is less than 1, the series con-

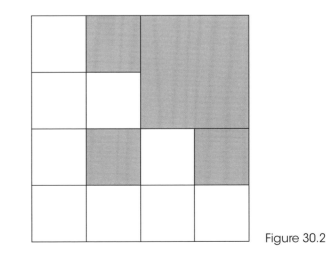

Figure 30.1

Figure 30.2

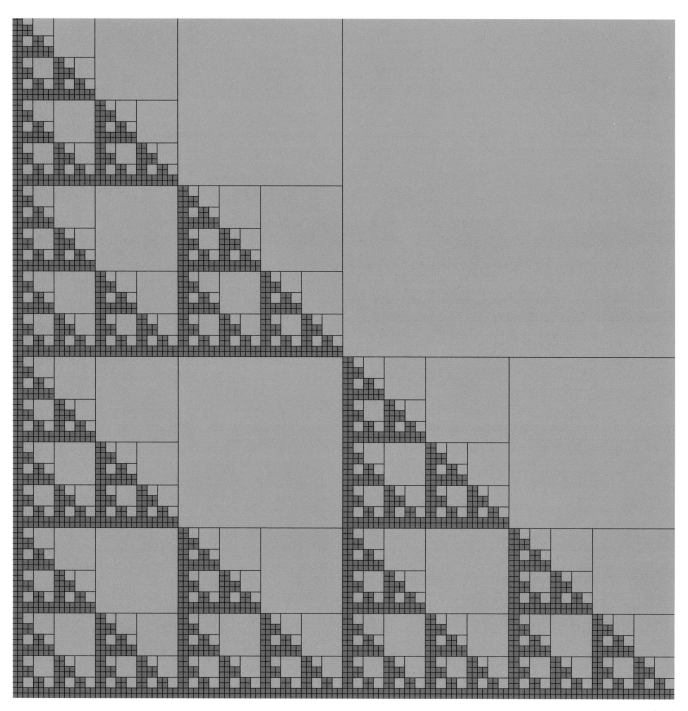

Plate 30. $\frac{3}{3} = \frac{4}{4}$

verges to the limit $1/(1-\tfrac{3}{4})=1/(\tfrac{1}{4})=4$ (see the appendix, page 177). Therefore, the total shaded area will be $\tfrac{1}{4}\times 4=1$—the area of the original square. In other words, after sufficiently many steps, the shaded part will nearly cover the entire original square, even though at each step we shaded only one-fourth of each square. Plate 30, *3/3 =4/4*, carries this process to its sixth stage; the shaded and unshaded squares are shown here in yellow and blue, respectively.

The infinite geometric series was already known to the Greeks; in fact, it is at the heart of Zeno's paradoxes (see again chapter 16). The Greeks knew that under certain conditions—specifically, when the common ratio is less than 1 in absolute value—the series seems to get closer and closer to a specific number. Today we call this number the *limit* of the series. The Greeks, however, could not perceive that the sum of the entire series is actually *equal* to the limit. A Greek mathematician would say only that with each additional shaded square, the combined area will get closer and closer to 1.

The geometric series is only the simplest of numerous series that have fascinated mathematicians over the ages. Infinite series and products were the

rage in the early and middle seventeenth century—just before the invention of the calculus—when the limit concept was not yet fully understood. All kinds of "strange" results followed from them, which left their discoverers with a sense of fascination. In 1668 the Danish-born mathematician Nicolaus Mercator (ca. 1620–1687) —one of many "minor mathematicians" who paved the way to the invention of the calculus—showed that the "infinite polynomial" (today we call it a power series) $x-x^2\!/2+x^3\!/3-x^4\!/4+\cdots$ converges to $\ln(1+x)$ whenever $-1<x\leq 1$. Put $x=1$ in the two expressions, and you get

$$\ln 2 = 1 - \frac{1}{2} + \frac{1}{3} - \frac{1}{4} + -\cdots,$$

a series as remarkable as the Gregory series (page 93). It is indeed puzzling why such a simple series—the sum of the reciprocals of the integers, taken with alternating signs—should have anything to do with ln 2 and by implication with *e*, a number intimately tied with the calculus. Unfortunately, there is no obvious way to illustrate such series geometrically; the geometric series is one of the few exceptions, so the blue and yellow squares of our plate will have to do.

31

The Harmonic Series

In the last chapter we saw that the series $1 - \frac{1}{2} + \frac{1}{3} - \frac{1}{4} + \cdots$ converges to $\ln 2$.

It is tempting to ask what will happen if we take the terms of this series in absolute value, that is, all positive. We then get the *harmonic series*, the sum of the reciprocals of the positive integers:

$$1 + \frac{1}{2} + \frac{1}{3} + \frac{1}{4} + \frac{1}{5} + \cdots$$

The name "harmonic" comes from the fact that a vibrating string produces not only one note but infinitely many higher notes, whose frequencies are 1, 2, 3, 4, 5, ... times the fundamental, or lowest, frequency. It is one more example of the influence that music, and musical terminology, has had on mathematics.

Since the Middle Ages it was known that the harmonic series diverges—its sum grows without bound as we add more and more terms, despite the fact that the terms themselves get smaller and smaller. But you would never guess this from watching the sum grow, because the *rate* of divergence is agonizingly slow. Some numbers will make this clear: the sum of the first thousand terms of the series is 7.485, rounded to the nearest thousandth; the sum of the first million terms is 14.357; the first billion terms, about 21; the first trillion terms, about 28. But to make the sum exceed, say, 100, we would have to add up a staggering 10^{43} terms (that's 1 followed by 43 zeros). To get an idea of just how large this number is, suppose we were to write down the series, term by term, on a long paper ribbon until its sum surpasses 100, allocating 1 cm for each term (this is actually an underestimate, since the terms will require more and more digits as we go along). The ribbon will then be 10^{43} cm long, which is about 10^{25} light-years. But the size of the observable universe is at present estimated at only 10^{11} light-years, so our ribbon would soon be running out of space to do the job! Yet if we could sum up the *entire* series—all its infinitely many terms—the sum would grow to infinity.

The divergence of the harmonic series was first proved by Nicole Oresme (ca. 1320–1382), a French theologian, economist, and mathematician. His proof is based on comparing the terms of the series with a second series in which the third and fourth terms are $\frac{1}{4} + \frac{1}{4}$ instead of $\frac{1}{3} + \frac{1}{4}$, the fifth, sixth, seventh, and eighth terms are $\frac{1}{8} + \frac{1}{8} + \frac{1}{8} + \frac{1}{8}$ instead of $\frac{1}{5} + \frac{1}{6} + \frac{1}{7} + \frac{1}{8}$, and so on. Since $\frac{1}{3} + \frac{1}{4} > \frac{1}{4} + \frac{1}{4}$, $\frac{1}{5} + \frac{1}{6} + \frac{1}{7} + \frac{1}{8} > \frac{1}{8} + \frac{1}{8} + \frac{1}{8} + \frac{1}{8}$, etc., we have

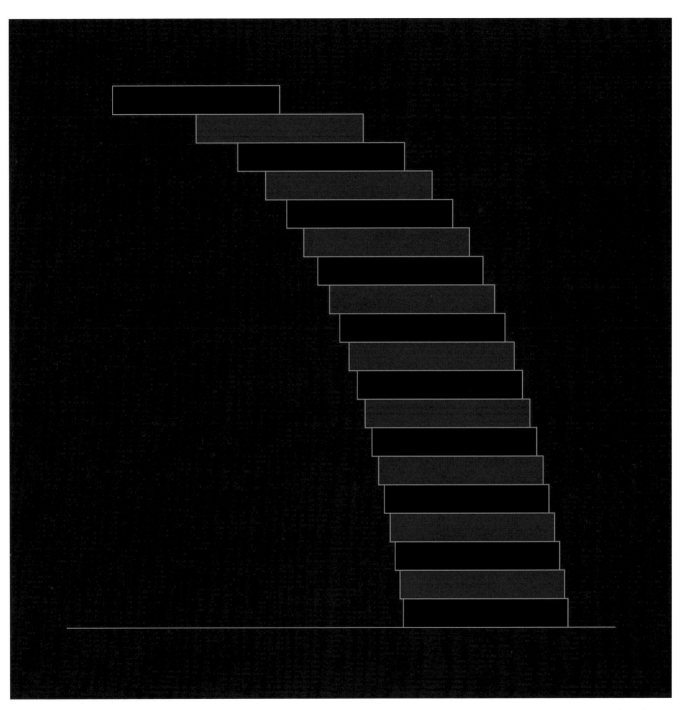

Plate 31. *A Dangerous Overhang*

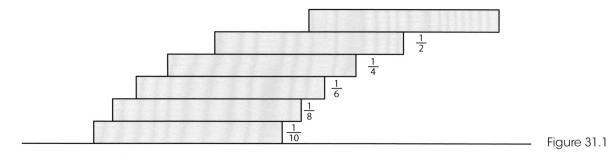

Figure 31.1

$$1+\frac{1}{2}+\frac{1}{3}+\frac{1}{4}+\frac{1}{5}+\frac{1}{6}+\frac{1}{7}+\frac{1}{8}+\cdots$$

$$>1+\frac{1}{2}+\underbrace{\frac{1}{4}+\frac{1}{4}}+\underbrace{\frac{1}{8}+\frac{1}{8}+\frac{1}{8}+\frac{1}{8}}+\cdots$$

$$=1+\frac{1}{2}+\frac{1}{2}+\frac{1}{2}+\cdots$$

Each group of terms after the first in this second series is equal to ½, so the series grows to infinity. And since the original series is *greater* than $1+\frac{1}{2}+$ $\frac{1}{2}+\frac{1}{2}+\frac{1}{2}+\cdots$, it too grows to infinity—it diverges.

The harmonic series gives rise to many surprises. For example, if we remove from it all the terms with composite denominators—leaving only prime-number denominators—the series will still diverge! This is rather remarkable, because the primes thin out as we go to higher numbers—they become ever more rare. And yet the sum of their reciprocals still diverges. On the other hand, the sum of the reciprocals of all *twin primes* (see page 45) is known to converge—although it is still an open question how many twin primes are there. So, until the issue is settled, we cannot say with unfailing certainty that this is an infinite series.

But perhaps the most amazing aspect of the harmonic series comes not from mathematics but from physics. Imagine stacking n identical domino tiles one on top of the other, but with each tile offset with respect to the one below it according to sequence ½, ¼, ⅙, ⅛, . . ., ½n (taking the length of each tile to be 1; see figure 31.1). The stack of dominoes will gradually curve and create an increasing overhang, and we would expect that its center of gravity will eventually extend beyond the bottom tile and cause the stack to collapse. Surprisingly, this does not happen: against all odds, the stack will survive intact, although just barely so (see plate 31). The overhang from stacking n dominos in this manner turns out to be ½$(1+\frac{1}{2}+\frac{1}{3}+\cdots+\frac{1}{n})$. And since this finite harmonic series diverges as $n \to \infty$, we could add on more and more tiles (starting at the bottom) and make the overhang as large as we please while the stack will maintain its equilibrium—though I wouldn't recommend standing under it on a rainy day![1]

NOTE:

1. See John Bryant and Chris Sangwin, *How Round is Your Circle: Where Engineering and Mathematics Meet* (Princeton, NJ: Princeton University Press, 2008), pp. 255–59.

32

Ceva's Theorem

Giovanni Ceva (1647–1734) was born in Milan and got his schooling in a Jesuit institute there. After completing his education in Pisa he was appointed professor of mathematics at the university of Mantua, where he stayed for the rest of his life. His work was in geometry and in hydraulics—not an unusual combination in seventeenth-century Europe, before academic specialization became the rule: Leonardo da Vinci, Galileo, and many other scientists were working on engineering problems as much as on art, mathematics, and physics.

The prevailing view at the time was that after Euclid, geometry became largely a closed subject, with nothing more of value to be discovered. Ceva disproved this view: in 1678 he discovered a new theorem that is remarkable for its simplicity.

Let ABC be any triangle, and let A', B', and C' be any points on the sides opposite of A, B, and C, respectively (figure 32.1). Ceva's theorem says that if the lines AA', BB', and CC' pass through one point, then

$$\frac{\overline{BA'}}{\overline{A'C}} \cdot \frac{\overline{CB'}}{\overline{B'A}} \cdot \frac{\overline{AC'}}{\overline{C'B}} = 1.$$

Note that this expression is a product of three ratios; we will simply call it the *triple product*. The con-

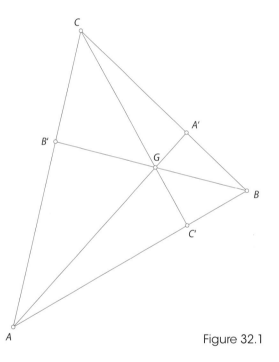

Figure 32.1

verse of Ceva's theorem is also true: if the triple product equals 1, the three lines are concurrent—they pass through one point.

In proving his theorem Ceva relied on mechanical principles, an approach that would not sit well with

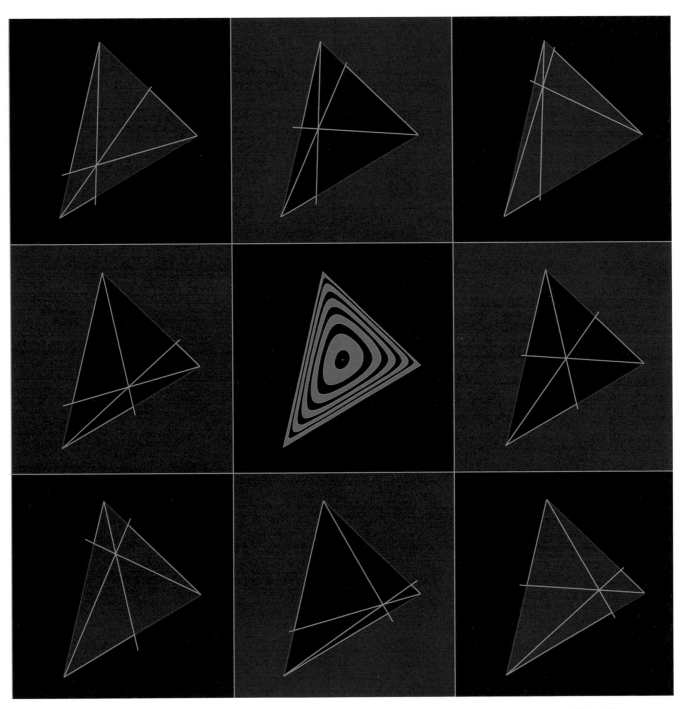

Plate 32. *Isocevas*

Euclid. Nevertheless, one cannot fail to admire the simplicity of the idea behind it—to think of *A*, *B*, and *C* as if they were physical objects with appropriate weights and then find their center of gravity *G*. Surprisingly, the actual values of these weights disappear from the final equation, leaving us with a purely geometric relation between the line segments in question. You will find Ceva's proof in the appendix.

But there is more to Ceva's theorem: if you look again at the triple product, you may be tempted to regard the letters as variables that can be multiplied and divided as if they were actual numbers. Then the six letters mutually cancel out, leaving the product to be equal to 1! To be sure, this is no more than a formal manipulation of symbols, but such formalism often plays a role in mathematics and may even hint at some deeper connections. Gottfried Wilhelm von Leibniz (1646–1716) used a similar symbolism to great advantage when he invented his version of the differential calculus.

If we rewrite the triple product as

$$\overline{BA'} \cdot \overline{CB'} \cdot \overline{AC'} = \overline{A'C} \cdot \overline{B'A} \cdot \overline{C'B},$$

we find that interchanging the primed letters with the unprimed ones does not affect the value of the product on each side—it remains invariant. This is shown in our illustration (plate 32): if you multiply together the lengths of the red-colored line segments and then do the same with the green-colored line segments, you end up with the same product.

The central triangle in the illustration shows several contour lines, each representing one value of the product and thus one position of *G*. In honor of Giovanni Ceva we will call them *isocevas*, in analogy with *isobars* and *isotherms*—lines connecting points of equal pressure or equal temperature on a weather map.

33

e

To the trio of special numbers we have met so far— $\sqrt{2}$, φ, and π—we now add a fourth number, e. This number, the base of natural logarithms, is of a more modern vintage than its ancient companions, tracing its origin to the seventeenth century. And unlike the others, it has its roots not in geometry, but in the world of business.

Financial matters have been of concern to people since the dawn of recorded history. A Babylonian clay tablet, dating back to about 1700 BCE and now in the Louvre, asks how long it would take for a sum of money, invested at 20 percent interest rate compounded annually, to double. This leads to the exponential equation $1.2^x = 2$, whose solution the tablet gives as about 3.787 years.[1]

The seventeenth century saw a renewed curiosity in the law of compound interest. Suppose $100 is invested at an interest rate of 5 percent compounded once a year. At the end of one year, this sum will grow to $100 \times 1.05 = \$105$. But suppose the bank compounds the interest *twice* a year, each time at *half* the nominal interest rate, that is, 2.5 percent. Since in the same year there are now two conversion periods, the accrued amount will grow to $100 \times 1.025^2 = \$105.06$, that is, 6 cents more than when interest is compounded annually.

Financial institutions use various methods of compounding money: annually, semiannually, quarterly, monthly, weekly, or even daily, in each case using an adjusted nominal interest rate: 0.05, $0.05/2 = 0.025$, $0.05/4 = 0.0125$, and so on. In each case the effective interest rate becomes smaller, but the money is compounded that much more often. Clearly we have here two opposing effects—a smaller interest rate but a higher number of compounding per year. Which will be the winner? Surprisingly, the two effects almost cancel out, with a very slight gain in favor of more frequent compounding, as shown in the following table:

$100 invested for 1 year at 5% interest rate compounded *n* times a year

Conversion period	n	0.05/n	Sum
Annually	1	0.05	$105
Semiannually	2	0.025	$105.06
Quarterly	4	0.00125	$105.09
Monthly	12	0.004166	$105.12
Weekly	52	0.0009615	$105.12
Daily	365	0.0001370	$105.13

Nothing in principle prevents us from compounding the sum even more often—every hour, every min-

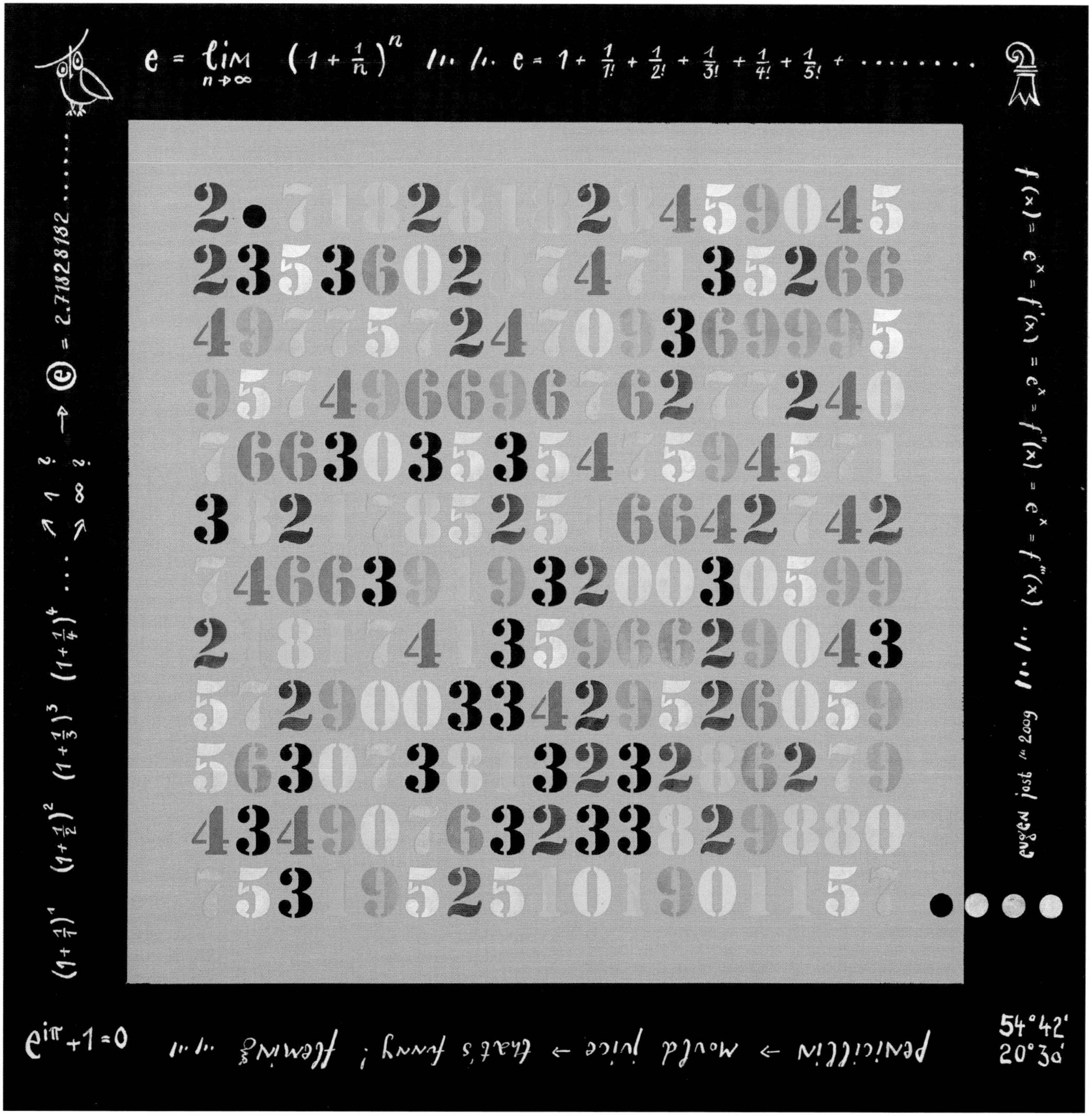

$$e = \lim_{n \to \infty} \left(1 + \frac{1}{n}\right)^n \qquad e = 1 + \frac{1}{1!} + \frac{1}{2!} + \frac{1}{3!} + \frac{1}{4!} + \frac{1}{5!} + \cdots$$

$$\left(1+\tfrac{1}{1}\right)^1 \quad \left(1+\tfrac{1}{2}\right)^2 \quad \left(1+\tfrac{1}{3}\right)^3 \quad \left(1+\tfrac{1}{4}\right)^4 \cdots$$

$$e = 2.7182182 \cdots$$

$$f(x) = e^x = f'(x) = e^x = f''(x) = e^x = f'''(x)$$

eugen jost "2009

$$e^{i\pi} + 1 = 0$$

penicillin → mould juice → that's funny! Fleming

54°42'
20°30'

Plate 33. *Euler's* e

ute, every second . . . every *instant*, resulting in a continuous compounding of the money. Someone in the sixteenth or early seventeenth century—we don't know who made the discovery or exactly when—noticed that as the interest is compounded more and more frequently, the accrued sum does not grow to riches, as one might expect, but seems to approach some limiting value. This limiting value comes from the formula $S = P(1 + r/n)^{nt}$ for the amount of money accrued when P dollars are invested for t years at an annual interest rate r compounded n times a year. If for convenience we replace r/n by $1/m$ in this formula, it becomes $S = P(1 + 1/m)^{rmt} = P[(1 + 1/m)^m]^{rt}$. As $n \to \infty$, so does m, but the expression $(1 + 1/m)^m$ will approach a limit, as shown in the following table:

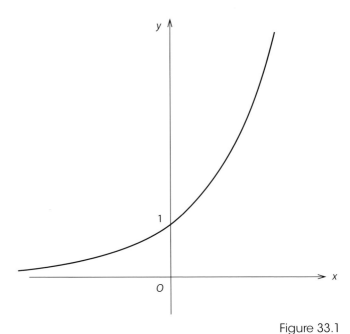

Figure 33.1

m	10	100	1,000	10,000	100,000
$(1 + 1/m)^m$	2.59374	2.70481	2.71692	2.71815	2.71827

This limit is about 2.71828 and is denoted by the letter e, a notation first used by Leonhard Euler in 1727.[2] Like $\sqrt{2}$, φ, and π, it is an irrational number. And like π, it is a *transcendental number*, meaning that it is not the solution of any polynomial equation with integer coefficients (see page 87).

It didn't take long before e showed up in other situations with no direct connection to finance. Foremost among them is the exponential function $y = e^x$, whose rate of change—its *derivative*, in the language of calculus—at any point x is always equal to e^x [that is, $(d/dx)e^x = e^x$]. It is this property that makes the exponential function so important in calculus—so much so that e is universally used as the natural base for all growth or decay processes, from the growth of a bacteria population to the disintegration of a radioactive substance. Figure 33.1 shows its graph.

Closely related to the exponential function are the expressions $(e^x + e^{-x})/2$ and $(e^x - e^{-x})/2$, collectively known as the *hyperbolic functions* and denoted by cosh x and sinh x, respectively. They bear some striking similarities to the more familiar trigonometric functions cos x and sin x; for example, in analogy with the trigonometric identity $\cos^2 x + \sin^2 x = 1$, we have the hyperbolic identity $\cosh^2 x - \sinh^2 x = 1$ (note, though, the presence of the minus sign). These analogies, however, are merely formal; in their graphs the two sets of functions are radically different, the hyperbolic functions lacking the periodicity of their trigonometric counterparts.

One of the outstanding problems that occupied the mathematical world in the years immediately following the invention of the calculus was to find the exact curve taken up by a chain of uniform thickness hanging freely under the force of gravity. To the eye a hanging chain looks very much like a parabola (see figure 33.2), but the Dutch physicist Christiaan Huygens (1629–1695) disproved this when he was just 17 years old. The mystery was solved in 1691 by three of the leading mathematicians of the time, who worked on it independently—Huygens himeslf (now 62), Leibniz, and Johann Bernoulii. To everyone's surprise, the shape turned out to be the graph of the hyperbolic cosine, $(e^x + e^{-x})/2$, which henceforth became known as the *catenary* (from the Latin *catena*, a chain). In our own time it has been immortalized in one of the world's most imposing architectural monuments, the Gateway Arch in St. Louis, Missouri, whose shape is exactly that of an inverted catenary.

Comparing the numerical values of $\sqrt{2}$, φ, e, and π—perhaps the four most famous numbers in mathematics—one cannot fail to notice how close they are on the number line, occupying less than three units of its infinite length. Why this is so no one knows; it remains one of the enduring mysteries of science.

Plate 33, *Euler's* e, gives the first 203 decimal places of this famous number—accurate enough for most practical applications, but still short of the exact value, which would require an infinite string of nonrepeating digits. In the margins there are several allusions to events that played a role in the history of *e* and the person most associated with it, Leonhard Euler: an owl ("Eule" in German); the Episcopal crosier on the flag of Euler's birthplace, the city of Basel; the latitude and longitude of Kö-

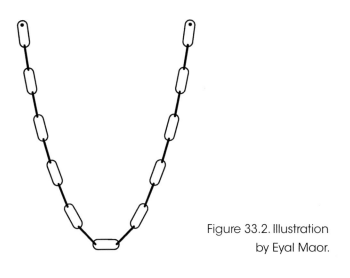

Figure 33.2. Illustration by Eyal Maor.

nigsberg (now Kaliningrad in Russia), whose seven bridges inspired Euler to solve a famous problem that marked the birth of graph theory; and an assortment of formulas associated with *e*.[3]

NOTES:

1. The tablet gives this answer in the peculiar base-60 Babylonian numeration system as (in modern notation) 3; 47,13,20, which stands for $3 + 47/60 + 13/60^2 + 20/60^3$, or about 3.7870.

2. Despite often-heard claims, it is unlikely that Euler chose the letter *e* because it is the initial of his own name.

3. For a more complete history of *e*, see Maor, *e: The Story of a Number*, of which this chapter is an excerpt.

34

Spira Mirabilis

Of the numerous curves we encounter in art, geometry, and nature, perhaps none can match the exquisite elegance of the logarithmic spiral, shown in plate 34.1. This famous curve appears, with remarkable precision, in the shape of a nautilus shell, in the horns of an antelope, and in the seed arrangements of a sunflower (see page 64, figure 20.1). It is also the ornamental motif of countless artistic designs, from antiquity to modern times. It was a favorite curve of the Dutch artist M. C. Escher (1898–1972), who used it in some of his most beautiful works.

The logarithmic spiral is best described by its polar equation $r = e^{a\theta}$, where r is the distance from the spiral's center O (the "pole") to any point P on the curve, θ is the angle between line OP and the x-axis, a is a constant that determines the spiral's rate of growth, and e is the base of natural logarithms. It follows from this equation that if θ is the sum of two angles θ_1 and θ_2, the radius will be the *product* of the corresponding radii: $r = e^{a(\theta_1 + \theta_2)} = e^{a\theta_1} \cdot e^{a\theta_2} = r_1 \cdot r_2$. Put differently, if we increase θ arithmetically (that is, in equal amounts), r will increase geometrically (in a constant ratio).

The many properties of the spiral all derive from this single feature. For example, a straight line from the pole O to any point on the spiral intercepts it at a constant angle α (figure 34.1).[1] It is for this reason that the curve is also known as an *equiangular spiral*. As a consequence, any sector with given angular width $\Delta\theta$ is similar to any other sector with the same angular width, regardless of how large or small it is. This is manifested beautifully in the nautilus shell (plate 34.2): the snail residing inside the shell gradually moves from one chamber to the next, slightly larger chamber, yet all chambers are exactly similar to one another: a single blueprint serves them all.

The logarithmic spiral has been known since ancient times, but it was the Swiss mathematician Jakob Bernoulli who discovered most of its properties. Bernoulli (1654–1705) was the senior member of an eminent dynasty of mathematicians, all hailing from the town of Basel. He was so enamored with the logarithmic spiral that he dubbed it *spira mirabilis* and ordered it to be engraved on his tombstone after his death. His wish was fulfilled—but not quite as he had intended: for some reason, the mason engraved a *linear* spiral instead of a logarithmic one (in a linear spiral the distance from the center increases arithmetically—that is, in equal amounts—as in the grooves of a vinyl record). The "wrong" spiral on

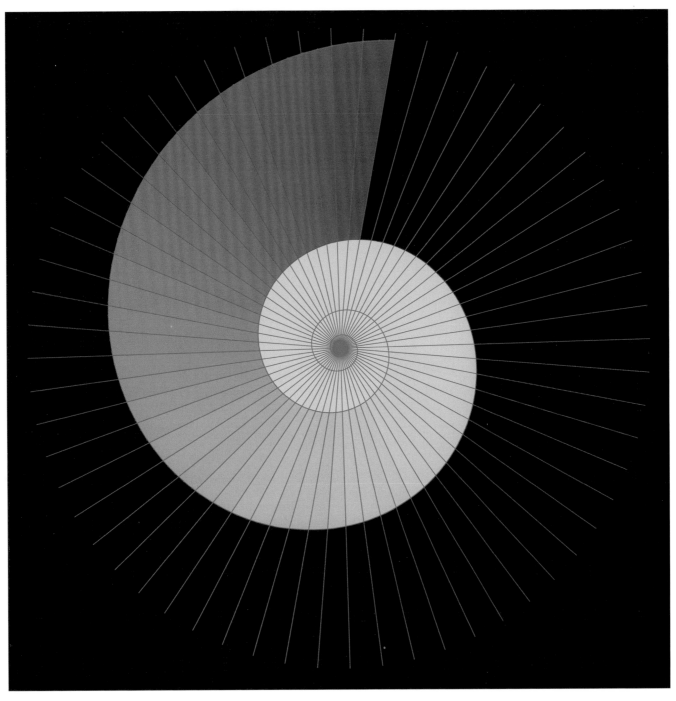

Plate 34.1. *Spira Mirabilis*

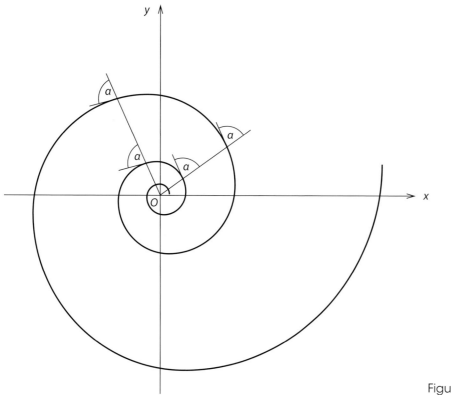

Figure 34.1

Bernoulli's headstone can still be seen at the cloisters of the Basel Münster, perched high on a steep hill overlooking the Rhine River (plate 34.3).

But if a wrong spiral was engraved on Bernoulli's tombstone, at least the inscription around it holds true: *Eadem mutata resurgo*—"though changed, I shall arise the same." The verse summarizes the many features of this unique curve: stretch it, rotate it, or invert it, it always stays the same.

NOTE:

1. This angle is determined by the constant a; in fact, $\alpha = \cot^{-1} a$. In the special case when $a = 0$, we have $\alpha = 90°$ and the spiral becomes the unit circle $r = e^0 = 1$. For negative values of a, the spiral changes its orientation from counterclockwise to clockwise as θ increases.

For more on the logarithmic spiral, see Maor, *e: The Story of a Number*, chapter 11.

Plate 34.2. *Nautilus Shell*

Plate 34.3. *Jakob Bernoulli's Tombstone*

35

The Cycloid

Rivaling the logarithmic spiral in elegance is the *cycloid*—the curve traced by a point on the rim of a circle that rolls along a straight line without slipping (figure 35.1). The cycloid is characterized by its arcs and cusps, each cusp marking the instant when the point on the wheel's rim reaches its lowest position and stays momentarily ar rest.

The cycloid has a rich history. In 1673 Christiaan Huygens, whom we've just met in connection with the catenary, solved one of the outstanding problems that had intrigued seventeenth-century scientists: to find the curve down which a particle, moving only under the force of gravity, will take the same amount of time to reach a given final point, regardless of the initial position of the particle. This problem is known as the *tautochrone* (from the Greek words meaning "the same time"). To his surprise, Huygens found that the curve is an arc of an inverted cycloid. He tried to capitalize on his discovery by constructing a clock whose pendulum was constrained to swing between two adjacent arcs of a cycloid, so that the period of oscillations would be independent of the amplitude (in an ordinary pendulum this condition holds only approximately). Unfortunately, although the theory behind it was

sound, the performance of Huygens's clock fell short of his expectations.

Shortly thereafter the cycloid made history again. In 1696 Johann Bernoulli (1667–1748), the younger brother of Jakob (of logarithmic spiral fame), posed this problem: to find the curve along which a particle, again subject only to the force of gravity, will slide down in the least amount of time. You might think this should be the straight line connecting the initial and final positions of the particle, but this is not so: depending on the path's curvature, the particle may accelerate faster at one point and slower at another, showing that the path of shortest *distance* between two points is not necessarily the path of shortest time.

Known as the *brachistochrone* ("shortest time"), this problem was attempted by some of the greatest minds of the seventeenth century. Among them was Galileo, who incorrectly thought the required path is an arc of a circle. In the end, five correct solutions were submitted in response to Johann Bernoulli's challenge—by Newton, Leibniz, L'Hospital (famous for a rule in calculus named after him), and the two Bernoulli brothers, who worked on the problem independently and used different methods. To their surprise, the

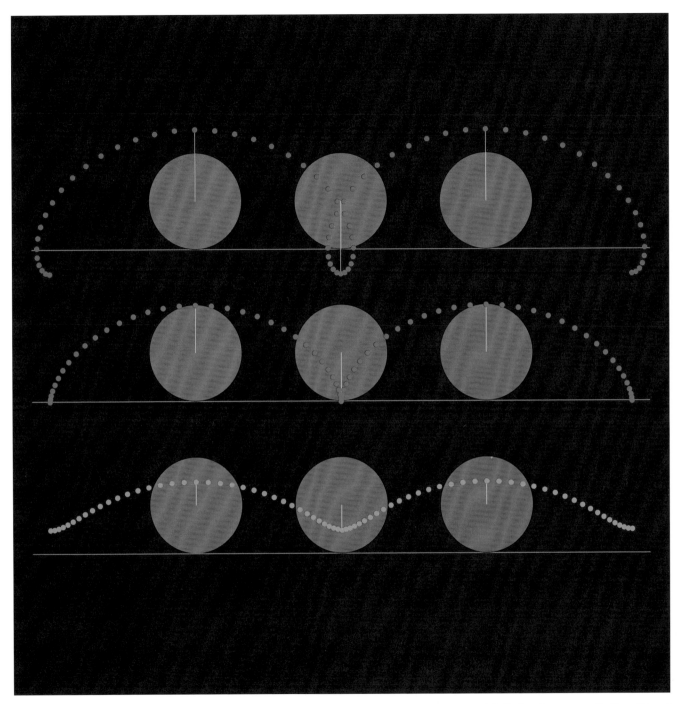

Plate 35. *Reflections on a Rolling Wheel*

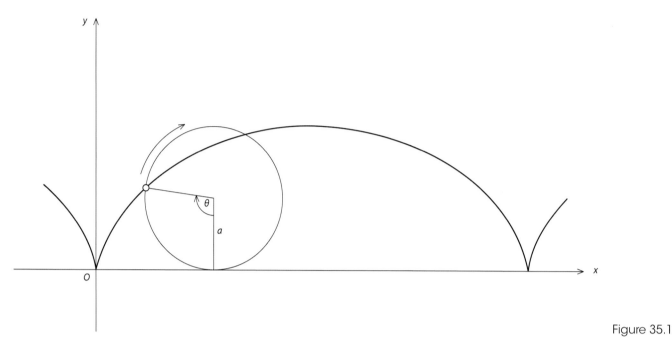

Figure 35.1

curve turned out to be an inverted cycloid—the same curve that solved the tautochrone problem. But instead of rejoicing in their success, the discovery embroiled the two brothers in a bitter priority dispute, resulting in a permanent rift between them.

The cycloid had some more surprises in store. Evangelista Torricelli (1608–1647), famous for his invention of the mercury barometer, is credited with finding the area under one arc of the cycloid: the area turned out to be $3\pi a^2$, where a is the radius of the generating circle. A few decades later Christopher Wren (1632–1723), London's venerable architect who rebuilt the city after the Great Fire of 1666, found that the length of each arc is $8a$; surprisingly, the constant π is not involved. This was one of the first successful *rectifications* of a curve—finding the arc length between two points on the curve. With the invention of calculus in the decade 1666–1676, problems like these could be solved routinely, but in

the early seventeenth century they presented a challenging task.

Plate 35, *Reflections on a Rolling Wheel*, shows the path of a luminous point attached to a rolling wheel at three different distances from the center: at top, the point is outside the wheel's rim (as on the flank of a railroad car wheel); at the middle, it is exactly on the rim; and at the bottom, inside of it. The top and bottom curves are called *prolate* and *curtate cycloids*, respectively, while the middle curve is the ordinary cycloid. You can see the curtate variant at night as the path traced by the reflector on a bicycle wheel as the cyclist moves forward.[1]

NOTE:

1. For a full history of the cycloid, see the article "The Helen of Geometry" by John Martin, *The College Mathematics Journal* (September 2009, pp. 17–27).

36

Epicycloids and Hypocycloids

Whereas the cycloid is generated by a point on the rim of a wheel rolling along a straight line, we may also consider a wheel rolling on the outside of a second, fixed wheel; the resulting curve is an *epicycloid* (from the Greek *epi*, meaning "over" or "above"). Or, we can let the wheel roll along the *inside* of a fixed wheel, generating a *hypocycloid* (*hypo* = "under"). The epicycloid and hypocycloid come in a great variety of shapes, depending on the ratio of the radii of the two wheels. Let the radii of the fixed and moving wheels be R and r, respectively. If R/r is a fraction in lowest terms, say m/n, the curve will have m cusps (corners), and it will be completely traced after n full rotations around the fixed wheel. If, however, R/r is not a fraction—if it is irrational—the curve will never close completely, although it will nearly close after many rotations. Figures 36.1 and 36.2 show the formation of a hypocycloid with a ratio $R/r = 5$ and an epicycloid with $R/r = 3$, respectively.

For some special values of R/r the ensuing curves can be something of a surprise. For example, when $R/r = 2$, the hypocycloid becomes a straight-line segment: each point on the rim of the rolling wheel will move back and forth along the diameter of the fixed wheel (figure 36.3). Thus, two circles with radii in

the ratio 2:1 can be used to draw a straight-line segment! In the nineteenth century this provided a potential solution to a problem that had vexed engineers for many years: how to convert the to-and-fro motion of the piston of a steam engine into a rotational motion of the wheels. It was one of many solutions proposed, but in the end it turned out to be impractical.

When $R/r = 4$, the hypocycloid becomes the star-shaped *astroid* (from the Greek *astron*, a star; see figure 36.4). This curve has some interesting properties of its own. Its perimeter is $6R$ (just as with the cycloid, this is independent of π), and the area enclosed by it is $3\pi R^2/8$, that is, three-eighths the area of the fixed circle.[1]

Imagine a line segment of fixed length with its endpoints resting on the *x*- and *y*-axes, like a ladder leaning against a wall. When the ladder is allowed to assume all possible positions, it describes a region bound by one-quarter of an astroid. This shows that a curve can be formed not only by a set of points lying on it, but also by a set of lines *tangent to it*. We will come back to this subject in chapter 40.

Turning now to the epicycloid, the case where the fixed and the moving wheels have the same radius ($R/r = 1$) is of particular interest: it results in a *cardi-*

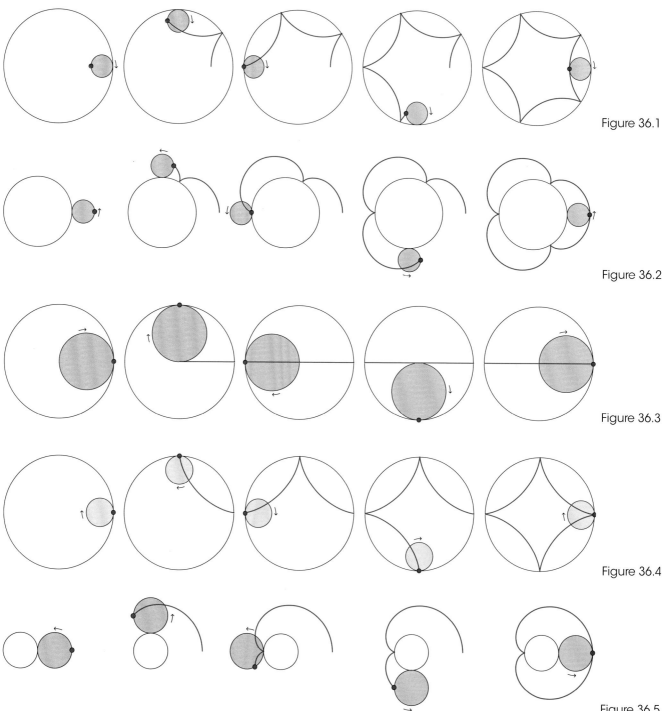

Figure 36.1

Figure 36.2

Figure 36.3

Figure 36.4

Figure 36.5

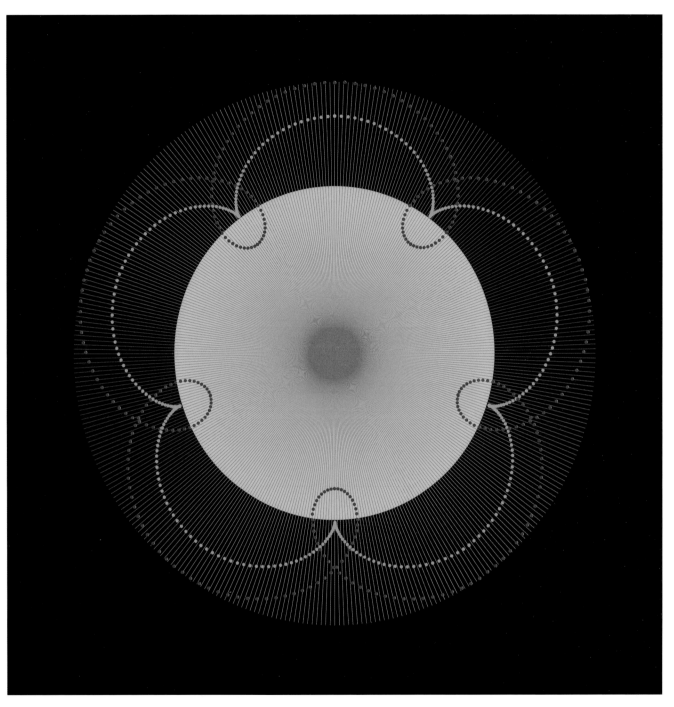

Plate 36. *Epicycloids*

oid, so called because of its heart-shaped form (figure 36.5). Its perimeter is $16R$ and its area is $6\pi R^2$.[2,3]

The Greek astronomer Claudius Ptolemaeus (commonly known as Ptolemy, ca. 85–165 CE), in an attempt to explain the occasional retrograde motion (east to west, instead of the usual west to east) of the planets, ascribed to them a complex path in which each planet moved along a small circle whose center moved around Earth in a much larger circle. The resulting orbit, an *epicycle*, has the shape of a coil wound around a circle. When even this model failed to account for the positions of the planets at any given time, more epicycles were added on top of the existing ones, making the system increasingly cumbersome. It was only when Johannes Kepler discovered that the planets move around the Sun in ellipses that the unwieldy epicycles became unnecessary and were laid to rest.

Plate 36 shows a five-looped epicycloid (in blue) and a prolate epicycloid (in red) similar to Ptolemy's planetary epicycles. In fact, this latter curve closely resembles the path of Venus against the backdrop of the fixed stars, as seen from Earth. This is due to an 8-year cycle during which Earth, Venus, and the Sun will be aligned almost perfectly five times. Surprisingly, 8 Earth years also coincide with 13 Venusian years, locking the two planets in an 8:13 celestial resonance and giving Fibonacci aficionados one more reason to celebrate!

NOTES:

1. We might mention in passing that the astroid has the unusual rectangular equation $x^{2/3} + y^{2/3} = R^{2/3}$.

2. For nice simulations of how these curves are generated, go to http://mathworld.wolfram.com/Hypocycloid.html and http://mathworld.wolfram.com/Epicycloid.html.

3. For more on the properties of epicycloids and hypocycloids, see Maor, *Trigonometric Delights*, chapter 7.

The Euler Line

Leonhard Euler (1707–1783) was arguably the most prolific mathematician of all time. Born in Basel, Switzerland, to a Calvinist minister, he studied with Johann Bernoulli and received his master's degree from the University of Basel at the age of 16. When two of Bernoulli's sons, Nicolas and Daniel (the latter famous for a law named after him in hydrodynamics), moved to St. Petersburg, Euler followed them and spent the next 14 years there. He then moved to Berlin to head the Prussian Academy of Sciences but returned to Russia in 1766 and never left again. His last years were beset by tragedies: the death of his wife, the loss of vision in both of his eyes, and a fire that destroyed his home and library (fortunately, most of his manuscripts were saved). None of these setbacks, however, slowed down his inexhaustible creativity: at the age of 70 he remarried, and although completely blind, he kept working to the very end, dictating his results to his assistants.

Euler's work covered nearly every area of mathematical research known in his time. He could move as easily from number theory to analysis as from geometry to physics and astronomy. He was also a great popularizer of science, writing numerous memoirs, letters, and books. His entire output, not yet fully published, is estimated to take up 70 volumes.

There are more formulas and theorems named after Euler than any other scientist in history. Among his most celebrated results are the formula $V - E + F = 2$ for the number of vertices V, the number of edges E, and the number of faces F of any simple polyhedron (a solid with planar faces and no holes), and the equation $e^{\pi i} + 1 = 0$ that relates the fundamental constants of arithmetic (0 and 1), of analysis (e), of geometry (π), and of complex numbers ($i = \sqrt{-1}$). This equation has often been hailed by popular writers as having no less than divine power, but mathematicians generally refrain from such mystical attributes.

We present here a lesser-known jewel of Euler, discovered by him in 1765: in any triangle, the *circumcenter* (the center of the circumcircle), the *centroid* (the intersection of the three medians, so called because it is the center of gravity of the triangle), and the *orthocenter* (the intersection of the three altitudes), all lie on one line, the *Euler line*. Moreover, if we denote the three points by O, G, and H, respectively, then $\overline{GH} = 2\overline{OG}$. This result looks so Euclidean, yet it had to wait for Euler to be discovered.

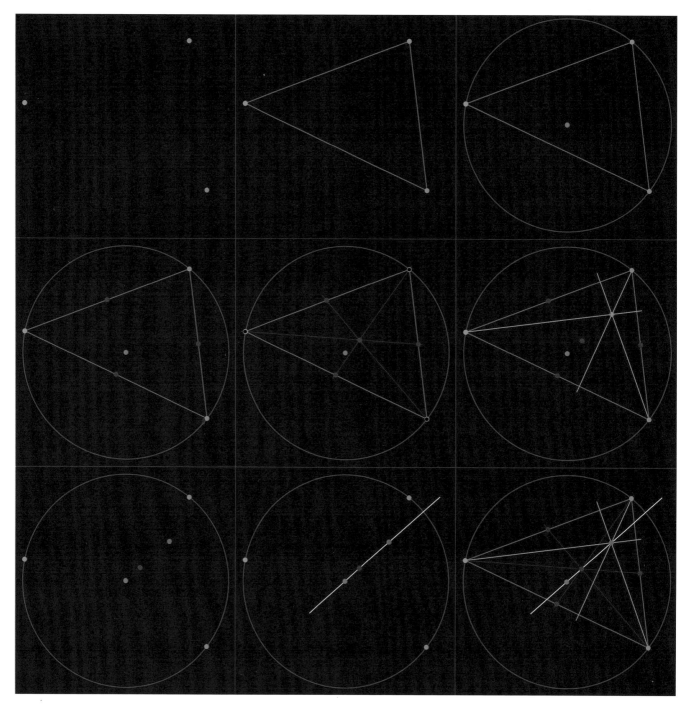

124

Plate 37. *Nine Points and Ten Lines*

Said the Canadian geometer H.S.M. Coxeter: "Some of his simplest discoveries are of such a nature that one can well imagine the ghost of Euclid saying, 'Why on earth didn't I think of that?'"[1]

Our illustration *Nine Points and Ten Lines* (plate 37) shows the point-by-point construction of Euler's line, beginning with the three points defining the triangle (marked in blue). The circumcenter *O*, the centroid *G*, and the orthocenter *H* are marked in green, red, and orange, respectively, and the Euler line, in yellow. We call this a construction without words, where the points and lines speak for themselves.

NOTE:

1. Coxeter, *Introduction to Geometry*, p. 17. For a proof of Euler's line theorem, see pp. 17–18.

38

Inversion

The circle can be a source of never-ending fascination. Consider a circle with center at O and radius 1. Two points P and Q on the same ray through O are called *inverses* of each other with respect to the circle if $\overline{OP} \cdot \overline{OQ} = 1$, or, equivalently, $\overline{OQ} = 1/\overline{OP}$ (figure 38.1). Either point can be thought of as the "image" of the other. If P is inside the circle (that is, $\overline{OP} < 1$), its image Q will lie outside ($\overline{OQ} > 1$), and vice versa. Therefore, the entire interior of the circle is "mapped" onto the exterior in a one-to-one correspondence. And, conversely, the interior is a miniature map of the world outside the circle. If $\overline{OP} = 1$, then $\overline{OQ} = 1$, so that points on the circle are mapped onto themselves.

But what about the center O? Where is its image point? If $\overline{OP} = 0$, then $\overline{OQ} = 1/\overline{OP} = 1/0$. But wait: didn't we learn in school that division by 0 is undefined? To go around this difficulty, we define the image of O under inversion to be *the point at infinity*. But where exactly is this point, and in what direction? The answer, metaphorically speaking, is: everywhere and nowhere! This "point" is not an ordinary point in the usual sense of the word; we introduce it only so as not to exclude the center from our discussion.

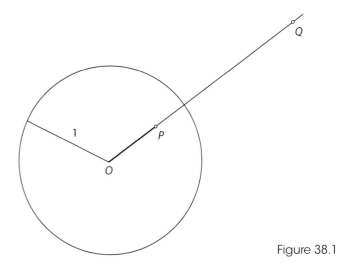

Figure 38.1

Since inversion maps points that are close to the center onto points far away from it, we expect that figures should be greatly distorted when subjected to inversion. Yet in spite of this, some figures do not change at all. Clearly, every ray through O is mapped on itself, and a circle with center at O and radius r becomes a circle with the same center and radius $1/r$. More surprising is the fact that a circle *through O* is mapped onto a straight line *not through O*, and vice

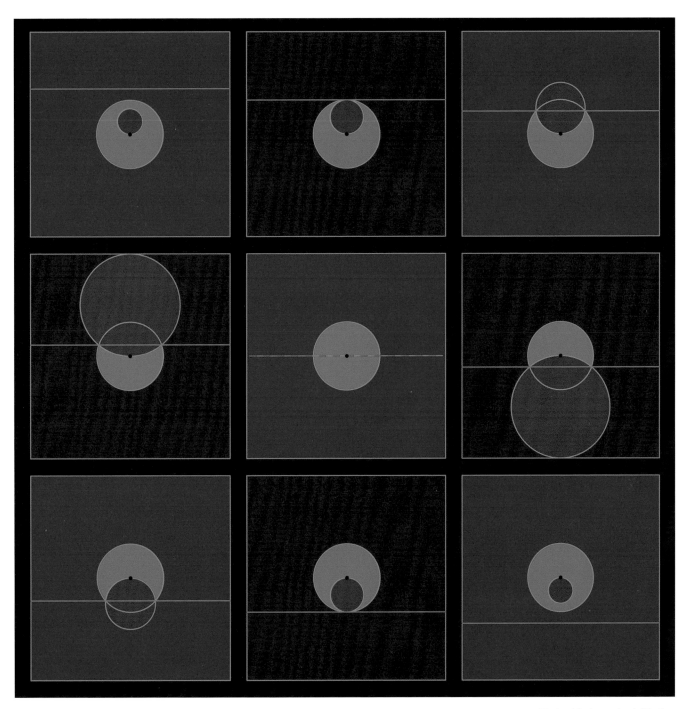

Plate 38. *Inverted Circles*

versa. Note that the two "endpoints" of the line are one and the same point, the point at infinity.

These relations are illustrated in plate 38. The circle of inversion, in light blue, is the same in all panels, while a red circle of varying size passes through the center of inversion, marked by a black dot. Depending on whether the red circle lies entirely inside the blue circle, touches it internally, or intersects it, its image line (in green) will pass outside the inversion circle, be tangent to it, or cut it at the two points where the circles intersect. The central panel illustrates the limiting case when the red circle is so large that it becomes a straight line *through* the center, in which case it coincides with its own image.

Inversion has many applications, of which we mention one here. In 1864 Charles-Nicolas Peaucellier (1832–1913), a French army officer, and Yom Tov Lipman Lipkin of Lithuania, the son of a famous rabbi, independently invented a mechanical linkage that could transform rectilinear motion into rotary motion (figure 38.2). It had previously been believed that this could not be done mechanically, but with the advent of the steam engine it became paramount to find a way to convert the to-and-fro motion of the piston into a rotary motion of the engine's wheels.

The Peaucellier inverter, as it became known, was one of many attempts at solving this problem. It consists of six rigid rods *OA*, *OC*, *AB*, *BC*, *CD*, and *DA* connected by linkages (points that allow free rotation) at *O*, *A*, *B*, *C*, and *D*, and such that $\overline{OA} = \overline{OC}$ and $\overline{AB} = \overline{BC} = \overline{CD} = \overline{DA}$. We have

$$\overline{OB} \cdot \overline{OD} = (\overline{OE} - \overline{BE}) \cdot (\overline{OE} + \overline{BE}) = \overline{OE}^2 - \overline{BE}^2$$
$$= (\overline{OA}^2 - \overline{AE}^2) - (\overline{AB}^2 - \overline{AE}^2) = \overline{OA}^2 - \overline{AB}^2.$$

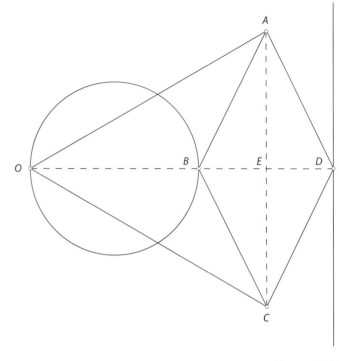

Figure 38.2

But $\overline{OA}$ and $\overline{AB}$ are given quantities, so the last expression is constant. Call this constant k^2 (it is positive because $\overline{OA} > \overline{AB}$). We then have $\overline{OB} \cdot \overline{OD} = k^2$, or

$$\frac{\overline{OB}}{k} \cdot \frac{\overline{OD}}{k} = 1,$$

showing that points *B* and *D* are inverses of each other in a circle (not shown in the figure) of radius *k* and center *O*. Consequently, when *D* moves along a straight line, its image *B* describes an arc of a circle through *O*, seemingly offering a solution to the rectilinear-to-circular conversion problem.[1]

Alas, it would never be possible to draw a complete circle with this device, as this would require us to trace the full length of the line, from one "endpoint" (the point at infinity) to the other. So the device was mainly a theoretical curiosity and became a favorite with nineteenth-century popularizers of science.[2]

NOTES:

1. A nice demonstration of the Peaucellier inverter can be found at http://mechanical-design-handbook.blogspot.com/2011/02/peaucellierlipkin-and-sarrus-straight.html.

2. A proof of some of the properties of inversion is found in the appendix. For a full discussion, see Coxeter, *Introduction to Geometry*, pp. 77–91.

39

Steiner's Porism

The first half of the nineteenth century saw a revival of interest in classical Euclidean geometry, in which figures are constructed with straightedge and compass and theorems are proved from a given set of axioms. This "synthetic," or "pure," geometry had by and large been thrown by the wayside with the invention of analytic geometry by Fermat and Descartes in the first half of the seventeenth century. Analytic geometry is based on the idea that every geometric problem could, at least in principle, be translated into the language of algebra as a set of equations, whose solution (or solutions) could then be translated back into geometry. This unification of algebra and geometry reached its high point with the invention of the differential and integral calculus by Newton and Leibniz in the decade 1666–1676, and it has remained one of the chief tools of mathematicians ever since. The renewed interest in synthetic geometry came, therefore, as a fresh breath of air to a subject that had by that time been considered out of fashion.

One of the chief protagonists in this revival was the Swiss geometer Jacob Steiner (1796–1863). Steiner did not learn how to read and write until he was 14, but after studying under the famous Swiss educator Heinrich Pestalozzi, he became completely dedicated to mathematics. Among his many beautiful theorems we bring here one that became known as *Steiner's porism*. Given two nonconcentric circles, one lying entirely inside the other, construct a series of secondary circles, each touching the circle preceding it in the sequence as well as the two original circles (see Figure 39.1). Will this *Steiner chain* close upon itself, so that the last circle in the chain coincides with the first? Steiner, in 1826, proved that if this happens for any particular choice of the initial circle of the chain, it will happen for *every* choice.

In view of the seeming absence of symmetry in the configuration, this result is rather surprising. Steiner devised a clever way of inverting the two original circles into a pair of *concentric* circles. As a result, the chain of secondary circles (now inverted) will occupy the space between the (inverted) given circles evenly, like the metal balls between the inner and outer rings of a ball bearing wheel; obviously these can be moved around in a cyclic manner without affecting the chain.

But that's not all: it turns out that the centers of the circles of the chain always lie on an ellipse (marked in red in figure 39.1), and the points of

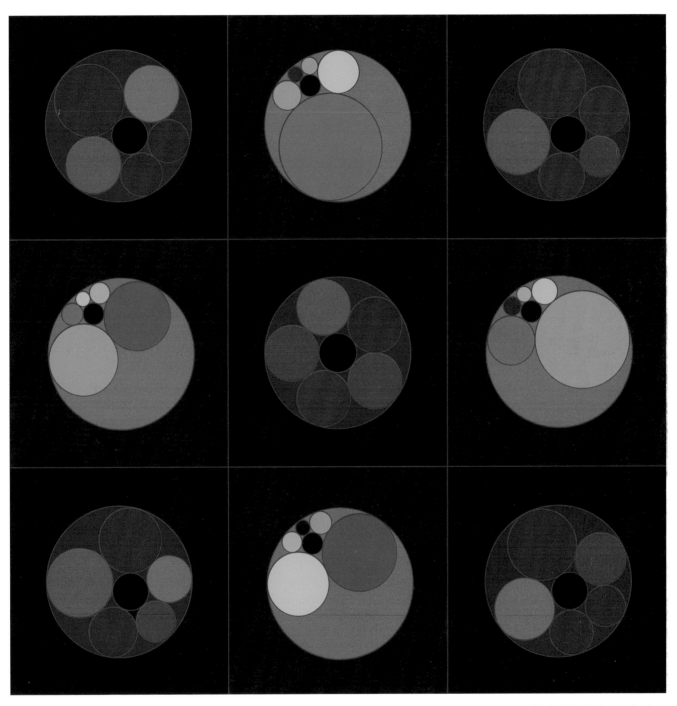

Plate 39. *Steiner's Porism*

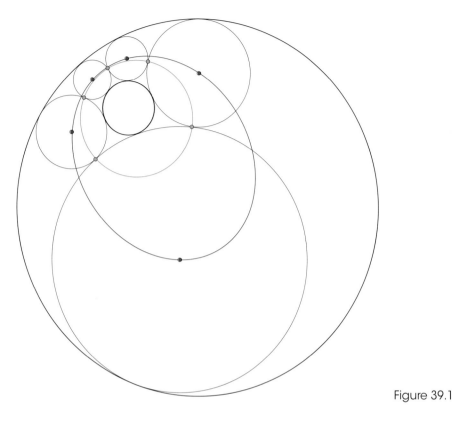

Figure 39.1

contact of adjacent circles lie on yet another circle (marked in green).[1]

Plate 39 illustrates several Steiner chains, each comprising five circles that touch an outer circle (alternately colored in blue and orange) and an inner black circle. The central panel shows this chain in its inverted, symmetric "ball-bearing" configuration.

As happens occasionally, a theorem that has been known in the West for many years turned out to have already been discovered earlier in the East. Steiner's porism is a case in point: a Japanese mathematician, Ajima Chokuyen (1732–1798), discovered it in 1784,

almost half a century before Steiner. An old Japanese tradition, going back to the seventeenth century, was to write a geometric problem on a wooden tablet, called *sangaku*, and hang it in a Buddhist temple or Shinto shrine for visitors to see. A fine example of Steiner's—or Chokuyen's—chain appeared on a sangaku at the Ushijima Chomeiji temple in Tokyo. The tablet no longer exists, but an image of it appeared in a book published about the same time as Steiner's discovery (see figure 39.2).[2]

It is somewhat of a mystery why this theorem became known as *Steiner's porism*. You will not find

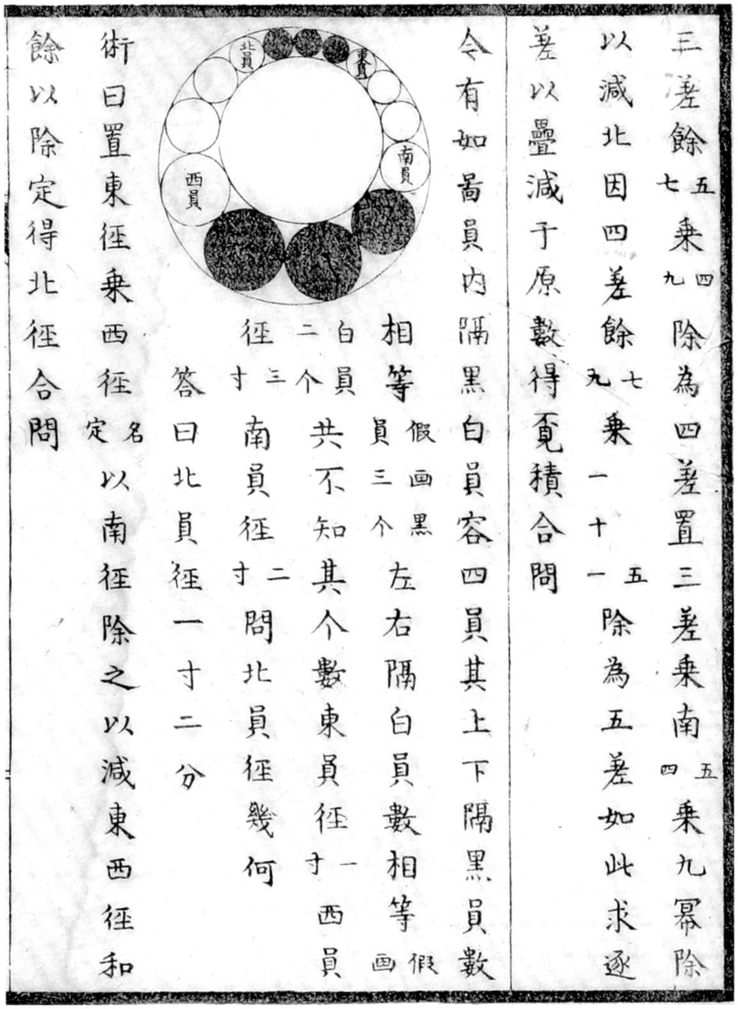

三差餘五七乘四九除為四四差置三差乘南五四乘九冪除
以減北因四差餘七九乘五一十一除為五差如此求逐
差以疊減于原數得覓積合問

今有如圖員內隔黑白員容四員其上下隔黑員數
相等假画黑員三个左右隔白員數相等假画白員二个
共不知其个數東員徑一寸西員
徑三寸南員徑二寸問北員徑幾何
答曰北員徑一寸二分
術曰置東徑乘西徑名定以南徑除之以減東西徑和
餘以除定得北徑合問

Figure 39.2. Courtesy of University of Aichi Education Library.

the word *porism* in your usual college dictionary, but the online Oxford English Dictionary defines it as follows: *In Euclidean geometry: a proposition arising during the investigation of some other propositions by immediate deduction from it*. Be that as it may, the theorem again reminds us that even the good old Euclidean geometry can still hold some surprises within it.

NOTES:

1. Steiner chains enjoy many additional properties. See http://en.wikipedia.org/wiki/Steiner_chain. For a proof of Steiner's porism, see Coxeter, *Introduction to Geometry*, p. 87.

2. See Hidetoshi and Rothman, *Sacred Mathematics: Japanese Temple Geometry*, p. 292.

40

Line Designs

Julius Plücker (1801–1868) is not a household name among present-day mathematicians, but in the nineteenth century he carved himself a niche in geometry where few others had ventured before. He realized that a curve need not be regarded as a set of points; it can just as well be described as a set of *tangent lines*. The idea was not entirely new. It had been known for more than a century that certain formal statements about points and lines remain valid when the words *point* and *line* are everywhere interchanged. For example, just as two points determine a unique line, so do two lines determine a unique point—their point of intersection.[1] This *principle of duality* became the centerpiece of a new kind of geometry, *projective geometry*, in which dual relations such as "two points determine a line" or "two lines determine a point" became the main focus, rather than metric properties such as the length of a line segment or the area of a polygon.

But while the principle of duality was well known in Plücker's time, he gave it a new formulation that placed the subject squarely in the realm of analytic geometry: he showed that a curve can be generated from a set of lines obeying a *line equation*, in much the same way as the traditional view of a curve as a set of points obeying a *point equation*. In other words,

lines can be used as building blocks of geometric figures just as much as points. Plate 40.1 shows a parabola generated entirely from its tangent lines; not a single point was used in its construction (the plate also shows the parabola's reflecting property, discussed earlier in chapter 29). Plate 40.2 goes even further, showing a Star of David–like design made of 21 line parabolas.

Plücker's career took him through strange twists. His major work in geometry was published in two volumes in 1828 and 1831, and it was in the second volume that he gave the analytic formulation of the principle of duality. Yet his work was not favorably received by the two leading geometers of the time, Jacob Steiner and Jean Victor Poncelet, whose synthetic geometry was more in line with the classical geometry of Euclid. It didn't help that Plücker's academic position at the University of Bonn was not in mathematics but in physics. This in itself was not unusual (Gauss held the position of director of the astronomical observatory at Göttingen), but it was used by Plücker's adversaries to claim that he was not a true physicist. To prove them wrong, he abandoned mathematics and for the next 18 years devoted himself to physics, making contributions in optics, magnetism, and spectroscopy. It was only

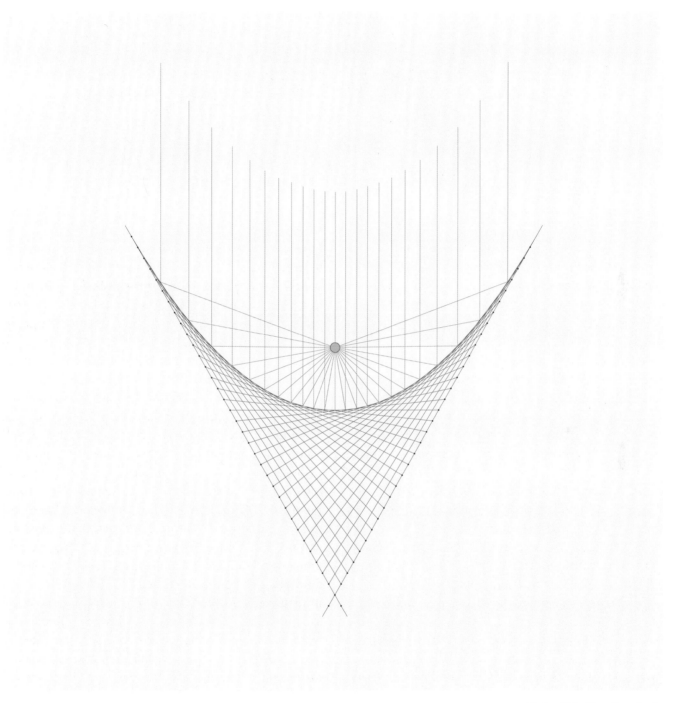

Plate 40.1. *Line Parabola*

Plate 40.2. *Line Design*

toward the end of his life that Plücker returned to his first love, geometry, where he made several more discoveries. After his death his work was completed by the influential German mathematician Felix Klein (1849–1925), who, like Plücker, was as much versed in algebra and geometry as he was in physics.[2]

NOTES:

1. If the lines are parallel, the point of intersection recedes to infinity and is known as a "vanishing point."

2. For a fuller discussion of line equations, see Maor, *The Pythagorean Theorem: A 4,000-Year History*, chapter 10.

41

The French Connection

There is a time-honored French tradition, going back at least to the sixteenth century, that calls for scientists to pursue a career in the military or civil service in parallel to their academic careers: François Viète, René Descartes, Jean Baptiste Joseph Fourier, and Victor Poncelet—to name but a few—all served as army officers, military engineers, or public administrators at various levels of government. This tradition is embodied by the lives of two early nineteenth-century French geometers, Charles Julien Brianchon and Jean Victor Poncelet.

Not much is known about the early life of Brianchon (1783?–1864); even his year of birth is in dispute, being given as either 1783 or 1785. At the age of 18 he enrolled at the prestigious École Polytechnique in Paris, where he studied under the geometer Gaspard Monge. While there he discovered a theorem that, in disguise, had already been found by another Frenchman, Blaise Pascal (1623–1662) more than one hundred fifty years earlier. Pascal's theorem, which he discovered when just 16 years old, says that if we inscribe a hexagon in an ellipse, the three points of intersection of pairs of opposite sides are collinear—they lie on one line (if two opposite sides are parallel, their point of intersection is "the point at infinity"). This is the exact dual of Brianchon's theorem: if we circumscribe a hexagon about

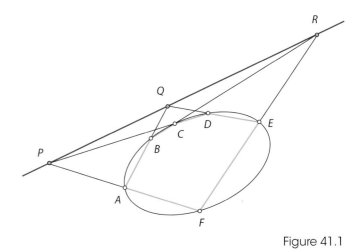

Figure 41.1

an ellipse, the three diagonals joining pairs of opposite vertices are concurrent—they pass through one point. The two theorems are illustrated in figures. 41.1 and 41.2.[1]

Later in life Brianchon joined the French artillery corps under the command of Napoleon, seeing action in Spain and Portugal. When Napoleon's campaign in the Iberian Peninsula ended in defeat, Brianchon got a teaching position at the Royal Artillery school in Vincennes. In his later years his scientific output declined, and he devoted himself almost entirely to teaching.

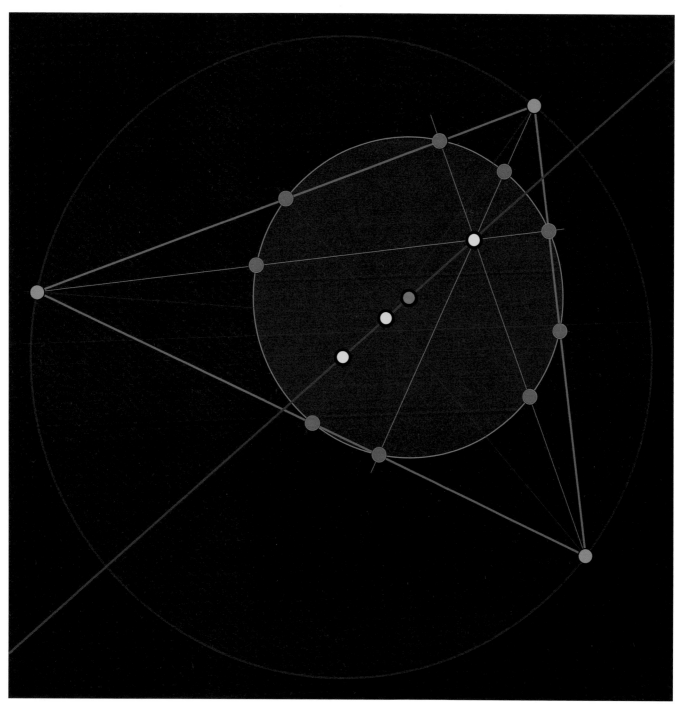

Plate 41. *French Connections*

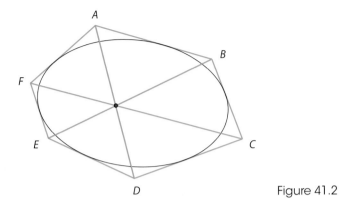

Figure 41.2

Poncelet (1788–1867) entered the École Polytechnique when he was 15. Like Brianchon, he then pursued a military career in the Corps of Engineers. When, in 1812, Napoleon launched his military campaign against Russia, Poncelet joined the ranks in the march eastward. In October of that year the tide turned against the French, and in the battle of Krasnoi he was injured and captured by the Russians. To keep his sanity during his long months in captivity and having no books to study from, Poncelet reconstructed from memory much of the mathematics he had learned at the Polytechnique and then used it to completely rewrite a branch of mathematics known as projective geometry, with the principle of duality as its cornerstone (see chapter 40).

In 1821 Brianchon and Poncelet together published the theorem for which they are best known, the *nine-point circle*: in any triangle, the feet of the altitudes, the midpoints of the sides, and the midpoints of the line segments joining the vertices to the orthocenter (the intersection point of the three altitudes), all lie on one circle. Furthermore, the center of this circle is halfway between the circumcenter and orthocenter, and its radius is one-half that of the circumcircle.[2] This is shown in plate 41, where the triangle and its circumcircle are colored in light and

dark blue, respectively. The nine points and the circle passing through them are shown in orange, with a green dot marking its center. The triangle's circumcenter, centroid, and orthocenter are in yellow, and the Euler line passing through them (see page 123) in red. Quite a lot of action is packed into this seemingly simple configuration!

The discovery of Brianchon's and Poncelet's theorem has other contenders, though: Karl Wilhelm Feuerbach (1800–1834) of Germany, at the age of 22, proved it for six of the nine points (the three median points and the three altitude feet); he discovered several additional properties associated with it, and the circle is often named after him. Olry Terquem (1782–1862), another French mathematician with a teaching career in the artillery corps, seems to have been the first to prove the theorem for all nine points, and he gave it the name *nine-point circle*. The occasional attribution of the theorem to Euler seems doubtful.[3]

NOTES:

1. The two theorems can be generalized to include any member of the conic sections—not just an ellipse—but this requires the introduction of points and lines at infinity, which are not part of Euclidean geometry.

For a proof of Brianchon's theorem, see Eves, *A Survey of Geometry*, pp. 143–44.

2. For a proof of the nine-point circle theorem, see Coxeter, *Introduction to Geometry*, pp. 18–19.

The nine-point circle has many other interesting features; see http://en.wikipedia.org/wiki/Nine-point_circle and http://mathworld.wolfram.com/Nine-PointCircle.html.

3. See the article "History of the Nine-Point Circle" by J. S. Mackay, *Proceedings of the Edinburgh Mathematical Society* [1892 (11), pp. 19–61], accessible on the Internet.

42

The Audible Made Visible

Ernst Florens Friedrich Chladni (1756–1827) is not among the giant names in the history of science, but his discoveries had a visual impact—quite literally. Born in Wittenberg, Germany, in the same year as Mozart, but outliving him by nearly forty years, Chladni was a rare combination of musician and physicist. In his book, *Discoveries in the Theory of Sound* (1787) he showed how the vibrations of sound-generating objects could be seen visually. He experimented with thin metal plates, over which a fine layer of sand was strewn. When the plate was excited by rubbing a violin bow against its edge, the sand arranged itself in beautiful geometric patterns of a high degree of symmetry. Chladni realized that these patterns—ridges on which the sand tends to accumulate—are *nodal lines*, lines along which the plate does not vibrate. This is analogous to the nodal points along a vibrating string, those points where the string divides itself into individual segments, each vibrating independently of the others.

Chladni went public with his discovery, demonstrating it before large audiences across Europe. Among his listeners was Napoleon—himself an enthusiastic supporter of scientific research—who rewarded Chladni with an honorarium of 6,000 francs.

Chladni evidently was a man of wide interests. Besides music and physics, he showed an interest in meteorology and speculated that meteorites—once thought to be the debris from volcanic eruptions—actually originated in outer space, a view that has since become the accepted fact. And in 1791—the year of Mozart's death—he invented a musical instrument, the *euphon*, consisting of metal bars and glass rods that were rubbed with a wet finger, somewhat similar to the glass harmonica for which Mozart wrote his Adagio and Rondo in C minor, K. 617. But it is his work on vibrating plates that earned Chladni his name in the history of science. He literally made sound visible.

Plate 42 shows nine Chladni figures generated in the physics lab of the Alte Kantonsschule in Aarau, Switzerland, under the supervision of our colleague Dr. Markus Meier. They were formed on triangular, square, and circular-shaped brass plates, over which a fine semolina powder was strewn. The vibrations were initiated by a wave generator that produced a nearly pure sine wave with frequencies ranging from 0.5 to 1.5 kHz (500–1500 cps). The electric signals were then converted to mechanical vibrations by a piezo element, and the ensuing patterns were photographed.

Plate 42. *Chladni Patterns*

43

Lissajous Figures

Another scientist who transformed sound into visual patterns was Jules Antoine Lissajous (1822–1880). Lissajous was professor of mathematics at the Lyceé Saint-Louis in Paris, where he studied all kinds of vibrations and waves. In 1855 he invented a simple optical device for analyzing compound vibrations. He attached small mirrors to the prongs of two tuning forks vibrating at right angles to each other. When a beam of light was aimed at one of the mirrors, it bounced off to the other mirror and from there to a screen, where it formed a two-dimensional pattern, the result of superimposing the two vibrations. This simple device—a forerunner of the modern oscilloscope—was a novelty in his time; up until then the study of sound depended entirely on the process of hearing, that is, on the human ear. Lissajous literally made it possible to "see sound."

Just as with snowflakes, Lissajous figures, as they came to be known, come in an infinite variety. Any change in the parameters of the two vibrations will drastically affect the ensuing figure. To see this, let

$$x = a \sin \omega_1 t, \quad y = b \sin (\omega_2 t + \varphi),$$

where a and b are the amplitudes of the two vibrations, ω_1 and ω_2 are their angular frequencies (in radians per second), and φ is the phase difference between them. As time progresses, a point P whose coordinates are (x, y) will describe a curve whose equation can be obtained by eliminating t between the two preceding equations. For example, if $\omega_1 = \omega_2$ and $\varphi = 0$ (the two vibrations being in tune and in phase), we get $y = (b/a)x$, the equation of a straight line through the origin; if, however, $\varphi = \pi/2$ (a 90° phase difference), we get $x^2/a^2 + y^2/b^2 = 1$, an ellipse.[1] If the phase difference slowly grows with time, this ellipse will continuously change its orientation and shape, passing (in the case $a = b$) from the circle $x^2 + y^2 = 1$ to the pair of lines $y = \pm x$. When the frequencies are unequal, the resulting figure becomes much more complex; if the ratio ω_1/ω_2 is an irrational number, the figure will never close—it will be aperiodic.

Due to their great variety, Lissajous figures became a favorite of nineteenth-century popular-science demonstrations. Their greatest expositor was none other than Lissajous himself, who—just like his predecessor, Chladni—made the lecture circuit all across Europe. And like Chladni, among Lissajous's audience was a distinguished guest—Emperor Napoleon, this time Napoleon III. Lissajous's experiments were shown at the Paris Universal Exposition

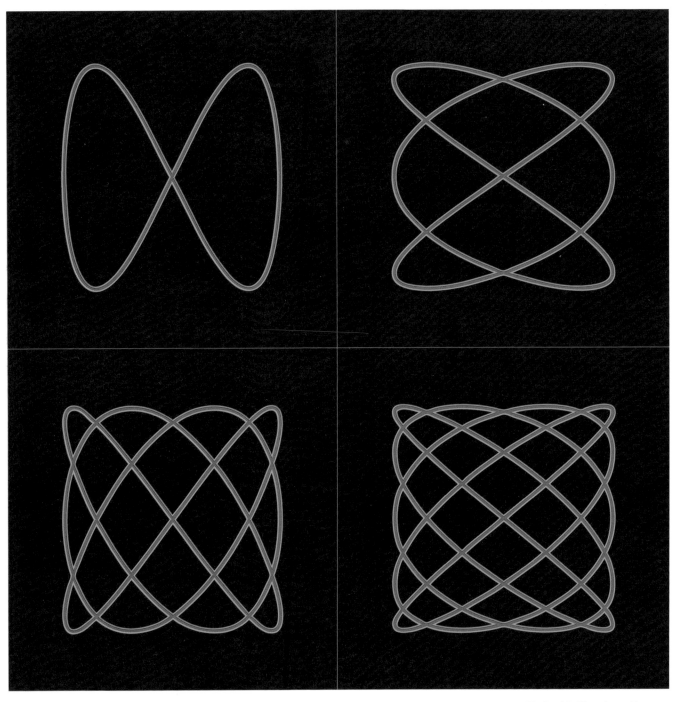

Plate 43. *Lissajous Figures*

in 1867, in recognition of which he was awarded the prestigious Lacaze Prize in 1873.

Our illustration (plate 43) shows four Lissajous figures with (going clockwise) frequency ratios of 1:2, 2:3, 3:4, and 4:5 and equal amplitudes. With graphing software you can create a virtually infinite variety of these figures, with arbitrary amplitudes, frequency ratios, and phase differences. It is fascinating to watch how even a small change in any of these parameters—and especially the frequency ratio—can drastically alter the ensuing figure. In years past, these figures were generated mechani-cally with a *harmonograph*—a device consisting of two coupled pendulums oscillating at right angles to each other; the frequencies could be varied by adjusting the length of each pendulum. You can still see a harmonograph at work at a few science museums, but the digital age has made it obsolete.

NOTE:

1. To see this, write $x = a \sin \omega_1 t$, $y = b \sin (\omega_1 t + \pi/2) = b \cos \omega_1 t$. Dividing the first equation by a and the second by b, squaring, and adding, we get $x^2/a^2 + y^2/b^2 = 1$.

44

Symmetry I

Few subjects have bridged the divide between the humanities and science more successfully than the concept of symmetry. Symmetry is as central to mathematics and physics as it is to the visual arts, architecture, music, and aesthetics. To the Greeks, symmetry meant a balance between the different parts of an object. Most Greek temples have a perfect *bilateral,* or *reflection,* symmetry: if you were looking at a mirror image of the Parthenon, you would hardly be able to distinguish it from the actual shrine. To the aesthetically minded Greeks, symmetry was synonymous with beauty and perfection; the human body was the ultimate example of such a perfection, and the Greeks realized it in their numerous statues and sculptures.

Mirror reflection is but one of several kinds of symmetry, indeed, the simplest one. In its broadest sense, symmetry is defined as the set of all transformations that, when acting on an object, leave that object *invariant,* or unchanged. These transformations may be reflections, rotations, or translations, and they may act on a physical object, a geometric figure, or a group of abstract symbols. An isosceles triangle, for example, will remain unchanged from its original orientation if we reflect it in the altitude through the top vertex; an *equilateral* triangle will

stay the same under a reflection in each of the three altitudes, as well as under 120°, 240°, and 360° rotations about its center (more about this in the next chapter).

In the nineteenth century these ideas led to the creation of a new branch of algebra known as *group theory*. Two names are associated with this development: the Norwegian Niels Henrik Abel (1802–1829) and the Frenchman Évariste Galois (1811–1832). The concept of a group came to them while attempting to find a formula for solving the general fifth-degree equation—the *quintic*—using only the elementary operations of addition, subtraction, multiplication, division, and root extraction, as with the familiar quadratic formulas for solving second-degree equations. As it happened, this turned out to be a pipe dream: Abel and Galois independently proved that no such formula exists for the general quintic, nor for the general equation of any higher degree (although *particular* cases of such equations may be solved in the manner described).

Tragically, both Abel and Galois died all too young: Abel of tuberculosis at the age of 27, Galois in a gun duel with a rival (ostensibly over a girl-friend, but more likely because of his antiroyalist political views in the years following the French

Plate 44. *Gothic Rose*

Revolution); he was just 20 years old. In their short lives they changed the course of mathematics, making it more formal, more abstract, and more general than ever before.

Plate 44, *Gothic Rose*, shows a rosette, a common motif on stained glass windows like those one can find at numerous places of worship. The circle at the center illustrates a fourfold rotation and reflection symmetry, while five of the remaining circles exhibit threefold rotation symmetries with or without reflection (if you disregard the inner details in some of them). The circle in the 10-o'clock position has the twofold rotation symmetry of the yin-yang icon. In the next chapter we will see how these symmetry patterns can be given a precise mathematical formulation.

45
Symmetry II

As we just saw, an equilateral triangle is endowed with six symmetry elements—three 120° rotations and three mirror reflections. Let the triangle be *ABC*, with vertex *A* at the top, followed clockwise by vertices *B* and *C*. Let us denote the six symmetry operations by letters: r_1, r_2, r_3 for the 120°, 240°, and 360° clockwise rotations and m_1, m_2, m_3 for the reflections in the altitudes through the top, lower-right, and lower-left vertices (henceforth we'll refer to these vertices as nos. 1, 2, and 3 rather than *A, B,* and *C,* because they keep changing their position as we rotate and reflect the triangle; thus the position-1 vertex is always at the top, 2 is at lower right, and 3 is at lower left, regardless of the positions of *A, B,* and *C*).

Now a 120° rotation changes triangle *ABC* to *CAB*, which is just a cyclic permutation of the letters *A, B* and *C,* so that *C* now occupies the 1-position, *A* occupies the 2-position, and *B* the 3-position (see figure 45.1). In a similar way, a 240° rotation transforms *ABC* into *BCA*, and a 360° rotation changes *ABC* into ... *ABC*—it brings the triangle back to its starting orientation. This, of course, should come as no surprise: a 360° rotation is the

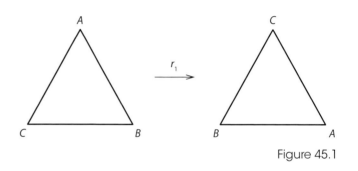

Figure 45.1

same as "doing nothing"—it leaves everything unchanged. To summarize,

$$r_1 : ABC \rightarrow CAB; \; r_2 : ABC \rightarrow BCA; \; r_3 : ABC \rightarrow ABC.$$

Turning now to the reflections, we can summarize them as follows:

$$m_1 : ABC \rightarrow ACB; \; m_2 : ABC \rightarrow CBA; \; m_3 : ABC \rightarrow BAC.$$

Note that in each reflection one letter stays put, while the other two switch positions—exactly what a mirror reflection does (figure 45.2).

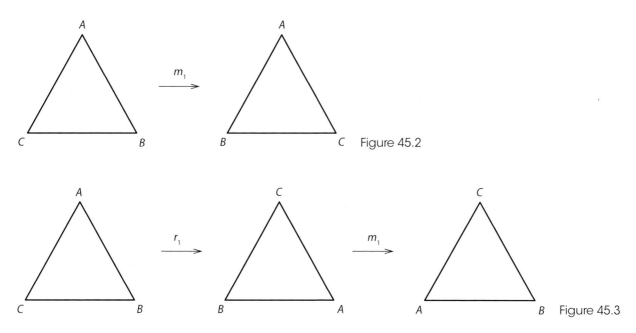

Figure 45.2

Figure 45.3

We are now ready to produce our trump card: we can follow any of the six symmetry operations with a second symmetry operation; for example, a 120° rotation followed by a reflection in the 1-position amounts to

$$ABC \rightarrow CAB \rightarrow CBA.$$

But *CBA* can be obtained directly from *ABC* by a reflection in the 2-position (in our case, vertex *B*; see figure 45.3). We call this combined application of two symmetry operations a *product* and denote it with a dot. And just as a product of two numbers gives us a third number, a product of two symmetry operations results in a third symmetry operation. In the example just given, this amounts to writing $r_1 \cdot m_1 = m_2$.

We have, in effect, created a kind of algebra of symmetry operations, similar in some ways to the ordinary algebra of numbers and variables we learn in school, but with one crucial difference: unlike numbers, the product of two symmetry operations is generally *noncommutative*—the order in which we perform the operations does matter. To see this, let us again find the product of r_1 and m_1, but this time in reverse order:

$$ABC \rightarrow ACB \rightarrow BAC.$$

That is to say, $m_1 \cdot r_1 = m_3$, whereas $r_1 \cdot m_1 = m_2$. Thus, $r_1 \cdot m_1 \neq m_1 \cdot r_1$.

If you are willing to spend a few more minutes on this exercise (it may remind you of your daily Sudoku), you can create a complete "multiplication

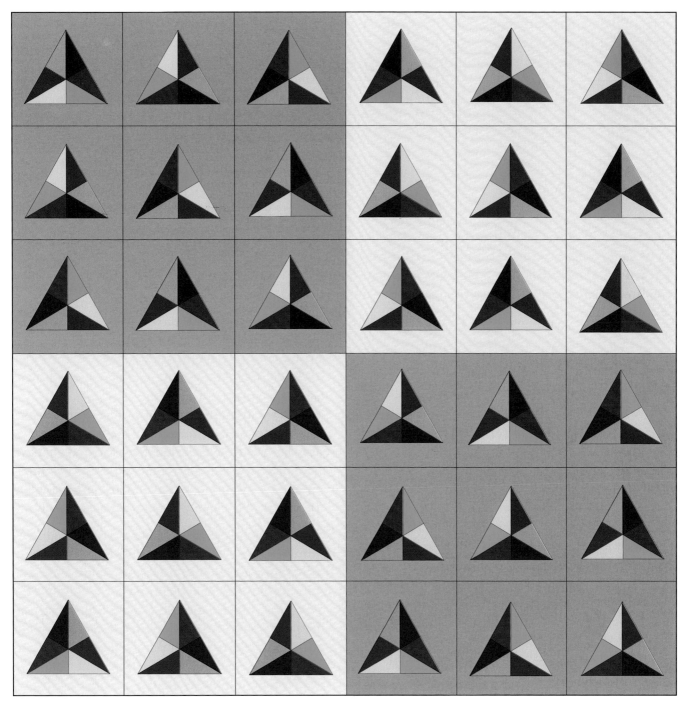

Plate 45. *Symmetry*

table" of the six symmetry operations of the equilateral triangle. Here it is:

·	r_1	r_2	r_3	m_1	m_2	m_3
r_1	r_2	r_3	r_1	m_2	m_3	m_1
r_2	r_3	r_1	r_2	m_3	m_1	m_2
r_3	r_1	r_2	r_3	m_1	m_2	m_3
m_1	m_3	m_2	m_1	r_3	r_2	r_1
m_2	m_1	m_3	m_2	r_1	r_3	r_2
m_3	m_2	m_1	m_3	r_2	r_1	r_3

Plate 45 shows all 36 entries of this table in a rainbow display of color. Note that the upper-left and lower-right quadrants (on gray background) consist of rotations alone, while the remaining two quadrants are pure reflections. We can summarize this in a miniature table:

·	r	m
r	r	m
m	m	r

Here, r and m stand for rotations and reflections of any kind. As this table shows, a succession of two rotations or two reflections always results in a rotation, whereas a rotation and a reflection result in a reflection. You can convince yourself of this when looking in the mirror: what you see is not yourself but a mirror image of yourself; if you stretch out your right hand, your image will respond with its left hand! To see a true image of yourself, you have to look at the intersection of two mirrors at right angles to each other; you will see yourself as someone else sees you.

Returning to the triangle's multiplication table, four features about this table are worth noting:

1. The table is "closed" in the sense that no matter which two symmetry operations you choose to multiply, their product will again be one of the six symmetry operations: you can never go outside this set.

2. Among the members of our set there is one element that has the effect of "doing nothing." This, of course, reminds us of the number 1 in ordinary multiplication: $1 \cdot a = a$ for any number a. And indeed, this particular element, called the *unit* or *identity* element, is r_3, the 360° rotation, since it leaves the triangle in its original position.

3. Just as with numbers, any multiplication in our set can be undone; that is, every element has an *inverse*, whose product with the original element results in the identity element. For example, the inverse of r_1 is r_2, because $r_1 \cdot r_2 = r_3$. The inverse of any reflection is the same reflection again, since the mirror image of a mirror image is the original image: a reflection is its own inverse.

4. If you multiply together three elements, the order of grouping does not matter; symbolically, $a \cdot (b \cdot c) = (a \cdot b) \cdot c$. This is the *associative law*, with which we are familiar from ordinary arithmetic. You can convince yourself of the validity of this law by trying a few examples (or, if you have the patience, all $432 = 2 \cdot 6^3$ possible combinations); for example, $r_1 \cdot (m_3 \cdot r_2) = r_1 \cdot m_1 = m_2$, while $(r_1 \cdot m_3) \cdot r_2 = m_1 \cdot r_2 = m_2$: you get the same answer.

Any collection of objects—whatever their nature—that fulfills these four requirements is called a *group*. It was this concept that Abel and Galois independently introduced in their quest for a formula for solving the general quintic (Galois suggested the name). For more than a hundred years the group concept was regarded as a purely abstract creation, devoid of any practical applications. But with the rise of modern physics in the twentieth century, group theory suddenly assumed central stage in nearly every branch of science, from crystallography and quantum mechanics to relativity and particle physics.

One of the great accomplishments of modern mathematics was achieved between 1995 and 2004 with the successful classification of all *finite simple groups*—groups with a finite number of elements from which all other groups can be created, in much the same way as any integer can be created by multiplying together its prime factors. This achievement, the result of a collaboration of more than 100 mathematicians, has been compared in its importance to the discovery half a century earlier of the DNA double helix and the genetic code.

46

The Reuleaux Triangle

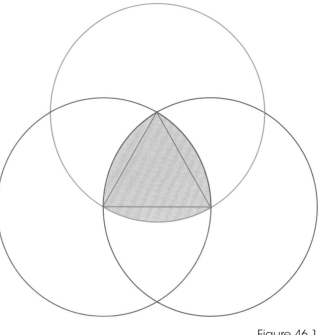

Franz Reuleaux (1829–1905), notwithstanding his French-sounding name, was a German scientist and engineer who is regarded as the founder of modern kinematics and machine design. He was born to a family of mechanical engineers and machine builders, an environment that nourished his future interests. He got his formal education at the Karlsruhe Polytechnic School and held his first academic appointment with the Swiss Federal Technical Institute (ETH) in Zurich. In 1864 he became professor at the Royal Industrial Academy in Berlin (later the Royal Technical College), being active as an educator, industrial scientist, and consultant. His views on various technical matters had considerable influence on the subsequent growth of German industry. But during the last years of his life his reputation declined: he was criticized for basing his ideas entirely on kinematic principles—that is, the purely geometric aspects of machinery—while ignoring the even more important dynamic effects of forces and torques.

Reuleaux's name is mainly remembered today for an idea that would lead to some unexpected practical uses. It all started with a simple geometric construction: with a compass, draw three identical circles, each of radius r and each passing through the

Figure 46.1

centers of the other two (figure 46.1). The overlapping central area is called the *Reuleaux triangle*. This simple-looking figure has some remarkable properties. For example, the width of the Reuleaux triangle is constant; that is, two parallel tangent lines to the

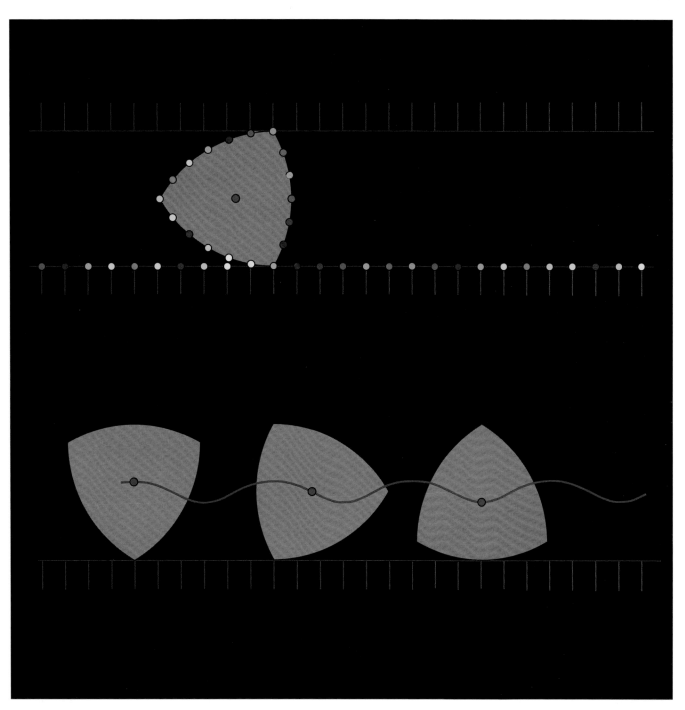

Plate 46. *Reuleaux Triangle*

triangle are always separated by a constant distance *r*, regardless of where we place them (note that one of these tangent lines will always pass through a vertex).

This feature, of course, reminds us of the circle, and indeed the two shapes have several properties in common. The circumference of the Reuleaux triangle is πr, the same as the circumference of a circle of *diameter r*. We can, therefore, regard the radius of the generating circles as the diameter of the Reuleaux triangle.[1] A Reuleaux triangle will always fit into a square of side *r*, a fact that inspired Harry James Watt (a descendant of the English inventor James Watt of steam engine fame) to issue a patent for a drill that could cut square holes (well, almost square: the corners would still be a bit rounded.)[2]

Because of its constant width, a Reuleaux triangle could be used as a wheel, at least in theory; but a ride on a vehicle using these "wheels" would be anything but smooth, as the axle would wobble up and down three times during each revolution (see plate 46). In Poul Anderson's science fiction story, *The Three-Cornered Wheel* (1963), inhabitants of an alien planet are using a noncircular, constant-width ob-

ject as their means of transportation, since the circle is regarded by them as a religious icon that should not be used for mundane purposes.

The Reuleaux triangle is but one of a large family of constant-breadth curves. It is known that all such curves with the same breadth *r* also have the same perimeter. Of all the curves in this set, the Reuleaux triangle has the smallest area, whereas the circle has the largest. This remarkable fact was proved in 1916 by the Austrian mathematician Wilhelm Blaschke (1885–1962). Thus, the circle and the Reuleaux triangle occupy the opposite ends of the family of constant-breadth curves.[3]

NOTES:

1. This analogy, however, does not extend to the triangle's area, which is $(r^2/2)(\pi - \sqrt{3})$, compared to that of a circle of diameter *r*, $\pi r^2/4$.

2. An animation of this drill can be seen at http://mathworld.wolfram.com/ReuleauxTriangle.html.

3. More on curves of constant breadth can be found in Rademacher and Toeplitz, *The Enjoyment of Mathematics*, chapter 25.

47

Pick's Theorem

Georg Alexander Pick (1859–1942) was born to a Jewish family in Vienna but spent most of his academic life in Prague. Pick began his career as assistant to some of the great names in mathematics and physics at the late nineteenth century, among them Ernst Mach (after whom the Mach number in aerodynamics is named) and Felix Klein, whose reform program in mathematics education would greatly influence subsequent generations of mathematicians. In 1892 Pick became professor at the German University of Prague, where 20 years later he would play a key role in offering a professorship to a young physicist by the name Albert Einstein; it would be Einstein's first full-time academic appointment. The two forged a close friendship, driven by their common passion for science and music.

Pick's work covered a wide range of subjects, including algebra, geometry, and analysis, but he is remembered today mainly for the theorem that bears his name, which he published in 1899. Imagine an infinite plane on which a uniform rectangular grid is superimposed. The points of intersection of this grid all have integer coordinates: $(0, 0)$, $(0, 1)$, $(1, 2)$, and so on. On this grid draw a polygon whose vertices all lie on the intersection points. Note that every side of this *lattice polygon* must connect at least

two intersection points, but it may contain more (for example, if the side is along a coordinate line). Pick discovered an unexpected relation between the area A of this polygon, the number of points M inside the polygon, and the total number of grid points on the sides N. Pick's theorem says that $A = M + \frac{1}{2}N - 1$.

Plate 47 shows a lattice polygon with 28 grid points (in red) and 185 interior points (in yellow). Pick's formula gives us the area of this polygon as $A = 185 + \frac{28}{2} - 1 = 198$ square units.

Pick's theorem received little attention during his lifetime, and even today few mathematics books mention it. It was mainly through a popular work, *Mathematical Snapshots* (1950) by the Polish mathematician Hugo Dyonizy Steinhaus, that the formula became known to a wider circle.[1]

If we rewrite Pick's formula as $2M + N - 2A = 2$, it bears a certain similarity to Euler's famous polyhedron formula $V + F - E = 2$ (see page 123). This similarity, however, is superficial: Pick's formula relates two pure numbers with a geometric quantity, area, while Euler's formula is a relation between three pure numbers; it does not tell us anything about the polyhedron's surface area or volume. Nevertheless, there is a subtle connection between them, which came to light shortly after Steinhaus made the theorem known.[2]

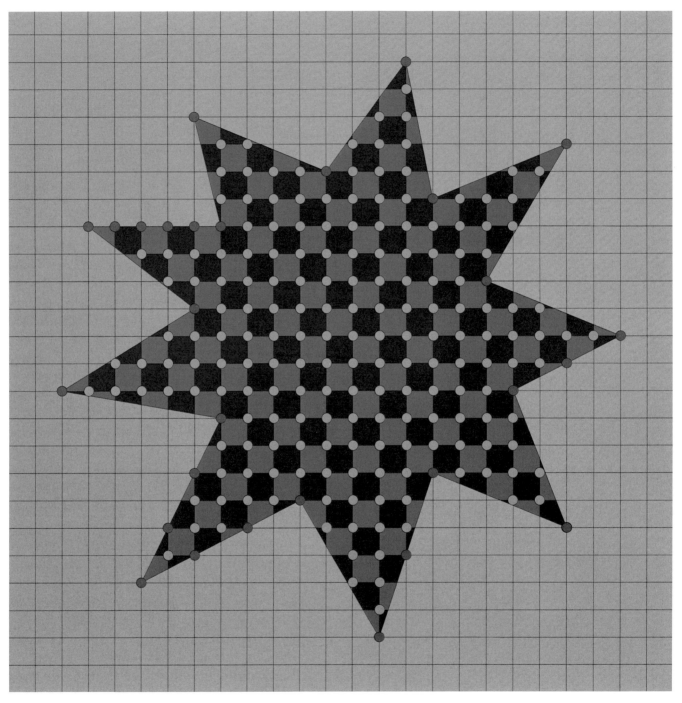

Plate 47. *Pick's Theorem*

Pick's later years were clouded by the political atmosphere of the 1920s and 1930s. Now retired, he hoped to live out his remaining years in Vienna, the place of his birth. But it was not to be: following the Anschluss—the Nazi occupation of Austria in 1938—he managed to go back to Prague, only to be rounded up there a year later when the Nazis invaded Czechoslovakia. Together with some one hundred fifty thousand fellow Jews, he was deported to the Theresienstadt concentration camp, where he died in 1942.

NOTES:

1. For a proof of Pick's theorem, see Coxeter, *Introduction to Geometry*, pp. 208–209.

2. See the following articles in *The American Mathematical Monthly*: "From Euler's Formula to Pick's Formula Using an Edge Theorem" by W. W. Funkenbusch [1974 (81), pp. 647–48]; "Pick's Theorem Revisited" by Dale E. Varberg [1985 (92), pp. 584–87]; and "Pick's Theorem" by Branko Grünbaum and G. C. Shephard [1993 (100)), pp. 150–61].

48

Morley's Theorem

Frank Morley (1860–1937) was an English-born mathematician who moved to the United States in 1887. He was professor at Johns Hopkins University from 1900 to 1928 and editor of the *American Journal of Mathematics* from 1900 to 1921. He published several books, including two on the theory of functions of complex variables (1893 and 1898), in which he introduced that subject—one of the highlights of nineteenth-century European mathematics—to American readers. He was known as an outstanding teacher who had no fewer than 45 doctoral students, an incredible number for any professor. He was also a gifted chess player and once beat the world champion Emmanuel Lasker.

But if Morley's name is remembered today, it is mainly for a theorem he discovered in 1899 but did not publish until 1929. Take any triangle *ABC* and draw its three pairs of *angle trisectors*, each pair originating at one vertex (see figure 48.1). Now consider the two trisectors adjacent to side *AB*; they meet at point *P*. Similarly, the trisectors adjacent to side *BC* meet at *Q*, and those adjacent to side *CA* meet at *R*. Morley's theorem says that points *P*, *Q*, and *R* form an equilateral triangle, appropriately called the *Morley triangle* and shown in yellow in the figure. This unexpected result came totally out of the blue; it

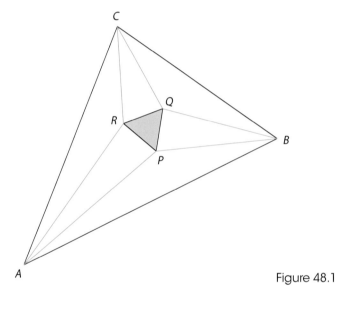

Figure 48.1

could have been included in Euclid's *Elements* if only Euclid had known about it—but he didn't, nor did anyone else for the next 2,200 years. It shows that even in classical geometry, surprises may still be awaiting us around the corner—or perhaps around the vertex!

The trisectors we have considered trisected the interior angles at *A*, *B*, and *C*, but Morley's theorem

Plate 48. *Morley's Theorem*

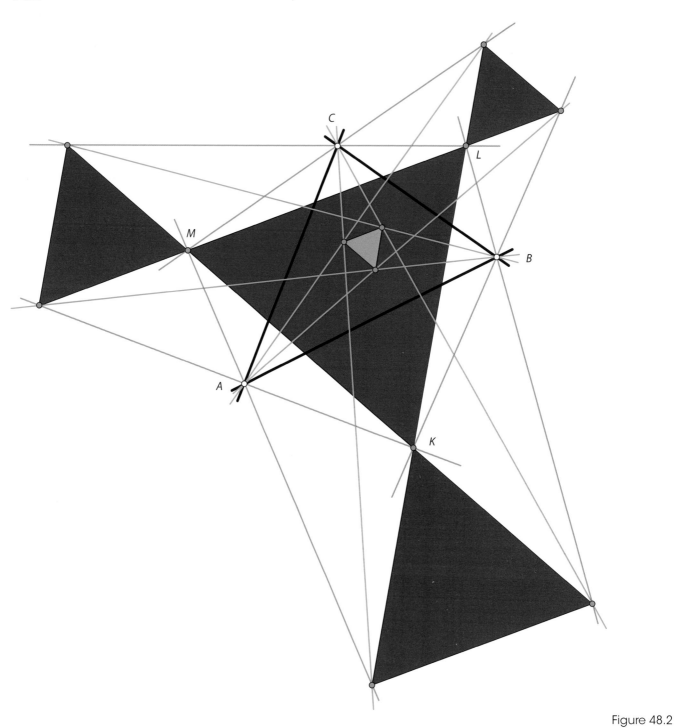

Figure 48.2

holds true for the *exterior* angles as well. Let the pairs of exterior trisectors adjacent to sides *AB*, *BC*, and *CA* meet at *K*, *L*, and *M*, respectively (figure 48.2); points *K*, *L*, and *M* again form an equilateral triangle, shown in purple in the figure. And that's not all: if we take the intersections of pairs of the remaining trisectors with the *extended* sides of this second Morley triangle, we get three more equilateral triangles (shown in red)! Considering that all five equilateral triangles were generated from a single, generic triangle, these results are truly remarkable.

One may wonder why this jewel of a theorem hadn't been discovered earlier. A possible explanation is that angle trisection is one of the three classical problems of antiquity that have no solution using only the Euclidean tools (see chapter 25). Consequently, problems involving angle trisection may have been regarded as of little interest and were, therefore, neglected.

As an additional bonus, the side of Morley's internal triangle is equal to $8r \sin(A/3) \sin(B/3) \sin(C/3)$, where r is the circumradius of the original triangle and A, B, C are its angles.

Plate 48 is a whimsical play on Morley's theorem, in which three key events of his life are alluded to. It shouldn't be too difficult to decode the meaning of the various symbols shown, so we leave this task to the reader.

NOTES:

1. For a proof of Morley's theorem, see Coxeter, *Introduction to Geometry*, pp. 23–25. Many alternative proofs and related results can be found at the Internet sites listed in the bibliography.

2. Figures 48.1 and 48.2 are color renditions based on the drawings in Wells, *The Penguin Dictionary of Curious and Interesting Geometry*, p. 155.

49

The Snowflake Curve

Niels Fabian Helge von Koch (1870–1924) was a Swedish mathematician who is remembered today chiefly, if not solely, for a curious curve he discovered in 1904. Take an equilateral triangle of unit side, divide each side into three equal parts, each of length ⅓, and delete the middle part (figure 49.1). Over the deleted part, construct the two sides of an equilateral triangle of side ⅓. This gives you a 12-sided Star of David shape, whose perimeter is ⁴⁄₃ that of the original triangle. Now repeat the process with each of the 12 new sides, resulting in a 48-sided shape whose perimeter is ⁴⁄₃ that of the previous perimeter, and thus $(⁴⁄₃)^2 = ¹⁶⁄₉$ of the original perimeter. Continuing in this way, the perimeter will increase by a factor of ⁴⁄₃ with each step. Since this factor is greater than 1, the perimeter keeps growing without bound and becomes infinite (more formally, $(⁴⁄₃)^n \to \infty$ as $n \to \infty$).

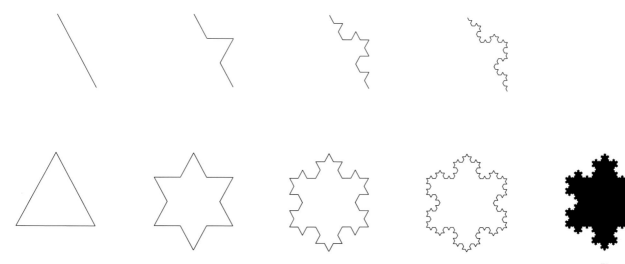

Figure 49.1

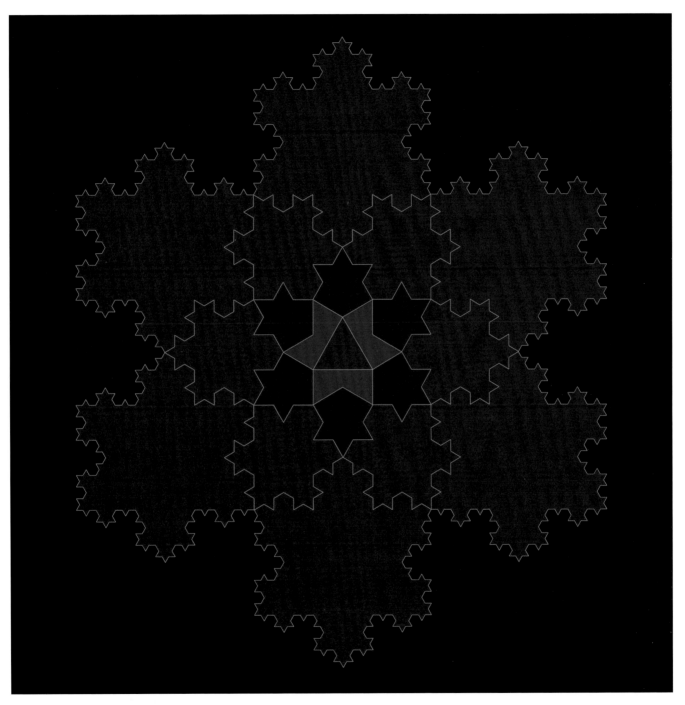

Plate 49. *Variations on a Snowflake Curve*

As you repeat the process again and again, the curve retains its overall Star of David shape but becomes increasingly crinkly. At the same time, the distance between any two points, no matter how close they are, becomes infinite, and, consequently, the curve has no tangent line anywhere. This, indeed, was the motivation behind its discovery: Koch attempted to show that there exist continuous curves with no tangent lines (curves that are nowhere differentiable, in the language of calculus). Such curves were a novelty in Koch's time, so much so that they were dubbed "pathological curves"—a hint to their strange behavior.

But what was strange yesterday has become the commonplace of today in the form of *fractals*—curves with the property that any small part of them looks exactly like the entire curve (mathematicians call this *self-similarity*). In fact, Koch's curve—also known as the *snowflake curve* due to its shape—was one of the first known fractals and certainly the first to make it into popular culture.[1]

One more feature of the Koch curve is worth noting: it encompasses a finite area. To see this, start again with the initial triangle of side 1. In the first step we add three small triangles, each of side $\frac{1}{3}$ and thus of area $\frac{1}{3}^2 = \frac{1}{9}$ of the area of the initial triangle. In the next step we add 12 new triangles, each of side $\frac{1}{9}$ and area $\frac{1}{9}^2 = \frac{1}{81}$ of the initial triangle. Continuing in this manner, we get the series

$$1 + \frac{3}{9} + \frac{12}{81} + \frac{48}{729} + \cdots = 1 + \frac{1}{3} + \frac{4}{3^3} + \frac{4^2}{3^5} + \cdots$$

$$= 1 + \frac{1}{3}[1 + 4/9 + (4/9)^2 + \cdots].$$

The expression inside the brackets is a geometric series with the common ratio $\frac{4}{9}$, so its sum is $1/(1 - \frac{4}{9}) = \frac{9}{5}$ (see page 51). The total area is therefore equal to $1 + \frac{1}{3} \cdot \frac{9}{5} = \frac{8}{5}$, or 1.6 times the area of the original triangle. So here we have a closed shape with a finite area but an infinite perimeter: you could never fence it off, as any fence would have infinitely many corners at every section of it. No wonder the Koch curve has become the darling of popularizers of mathematics.

Plate 49 is an artistic interpretation of Koch's curve, starting at the center with an equilateral triangle and a hexagram (Star of David) design but approaching the actual curve as we move toward the periphery.

NOTE:

1. The name *fractal* is credited to the French-born American mathematician Benoit Mandelbrot (1924–2010), who coined it in 1975.

Sierpinski's Triangle

Waclaw Franciszek Sierpinski (1882–1969) was born in Warsaw. He entered the University of Warsaw in 1899, graduating in 1904; four years later he was appointed to the University of Lvov. He became interested in set theory after reading about a theorem that allows for a seemingly impossible situation. We learn in analytic geometry that any point in the plane can be uniquely specified by two numbers, its x- and y-coordinates. This has been the rock foundation of analysis ever since René Descartes invented his analytic geometry in 1637. Not so, said the theorem that Sierpinski came across: one number suffices. Intrigued, he wrote to the article's author, Tadeusz Banachiewicz, asking for an explanation. He got a reply of a single word: "Cantor," referring to the creator of modern set theory, Georg Cantor, the subject of our last chapter. Becoming hooked on the subject, Sierpinski started studying it

and gave the first-ever university course devoted entirely to set theory. The topic would occupy him for the rest of his life.[1]

As an offshoot of his research, Sierpinski in 1915 came up with a seemingly impossible geometric configuration: a triangle-like figure whose area is zero. Start with an equilateral triangle (figure 50.1), and remove from it the small central triangle formed by connecting the midpoints of the sides (shown in white in the figure). Now repeat the process with the three remaining black triangles, then repeat it again and again . . . forever. In the limit, what will be the area of all the black triangles?

Taking the area of the initial triangle to be 1, each of the three smaller black triangles has an area ¼, so their combined area is $3 \times ¼ = ¾$. In the next step we have nine black triangles, each with area ¹⁄₁₆, making the total area $9 \times ¹⁄₁₆ = (¾)^2$. Continuing in this man-

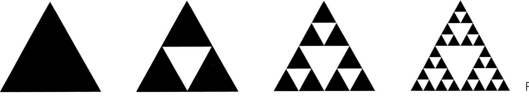

Figure 50.1

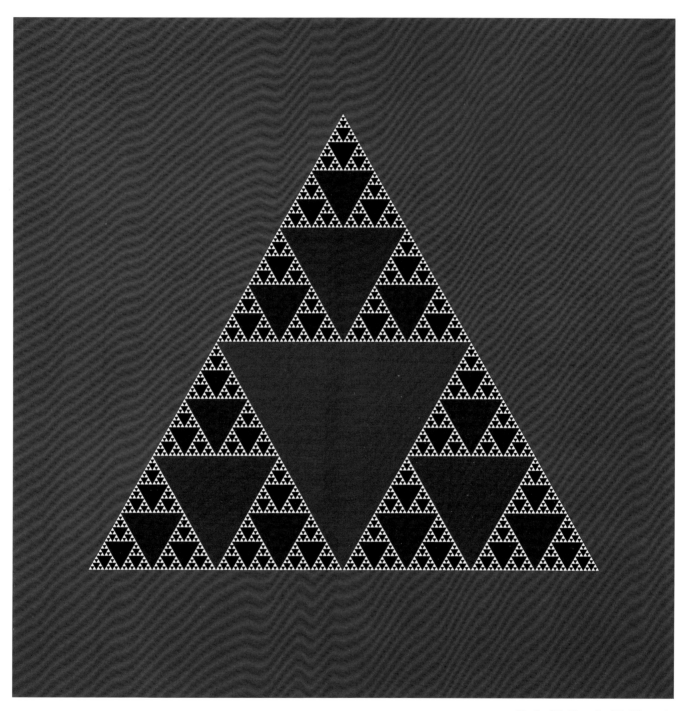

Plate 50. *Sierpinski's Triangle*

ner, the black areas follow the progression 1, ¾, $(¾)^2$, $(¾)^3$, ..., a geometric progression with the common ratio ¾. Since this ratio is less than 1, the terms of the progression tend to 0 as $n \to \infty$. So, eventually the original black triangle will become empty, despite the fact that at each stage we removed only ¼ of each black area.

On the other hand, the perimeters of the black triangles follow the sequence 3, 9⁄2, 27⁄4, 81⁄8, ..., a geometric progression with the common ratio 3⁄2. Since this ratio is greater than 1, the terms grow without bound as we keep removing more and more triangles, making the limiting perimeter infinite. That is to say, the limiting shape—known as *Sierpinski's triangle*—has zero area but infinite length! It shows again that when infinity comes into play, strange things always lurk around the corner.[2]

While we cannot go to infinity in the actual world, we can get pretty close with a computer, stopping only when the resolution reaches its limit at the pixel level. Plate 50 shows the seventh stage of Sierpinski's triangle.

NOTES:

1. This biographical sketch is based on the article "Waclaw Sierpinski" by John J. O'Connor and Edmund F. Robertson in *The MacTutor History of Mathematics Archive* (on the Web), listed in the bibliography.

2. The formation of Sierpinski's triangle can be viewed as an animation at http://en.wikipedia.org/wiki/Sierpinski_triangle.

51

Beyond Infinity

Can anything be larger than infinity? No, says common sense. But who is to say that common sense is always right—especially when it comes to the infinite, a world beyond our physical reach? "Infinity is a place where things happen that don't," an anonymous school pupil once said. Railroad tracks, though perfectly parallel, seem to meet far away on the horizon—at infinity; indeed, in projective geometry they *do* meet at infinity (see chapter 40). Yet when we try to reach that elusive point, it recedes from us just as fast as we approach it.

Again, when comparing two infinite sets, who is to tell which is the larger? There seem to be twice as many counting numbers as there are *even* numbers, and yet we can match every counting number with its double, showing that the two sets are just as large. And certainly there should be more points on a long line segment than on a short one, yet they both contain the same number of points—the same, in fact, as the entire number line (see figure 51.1). Strange, indeed, is the world of infinity.

Enter Georg Cantor (1845–1918). Born in St. Petersburg, Russia, his family moved to Germany when he was 11. In 1869 he settled in the city of Halle, at whose university he would spend most of

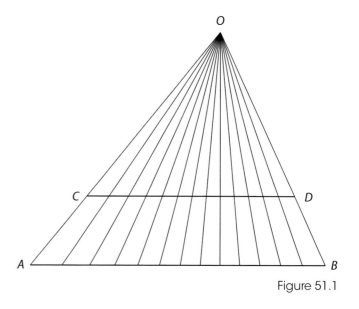

Figure 51.1

his productive years. Beginning in 1874, Cantor published a series of articles that at once put the concept of infinity on its head. To begin with, he insisted that an infinite set should be regarded in its entirety, as one whole. This ran smack against the accepted notion—going back to the Greeks—that infinity can be thought of only as a process, never as a

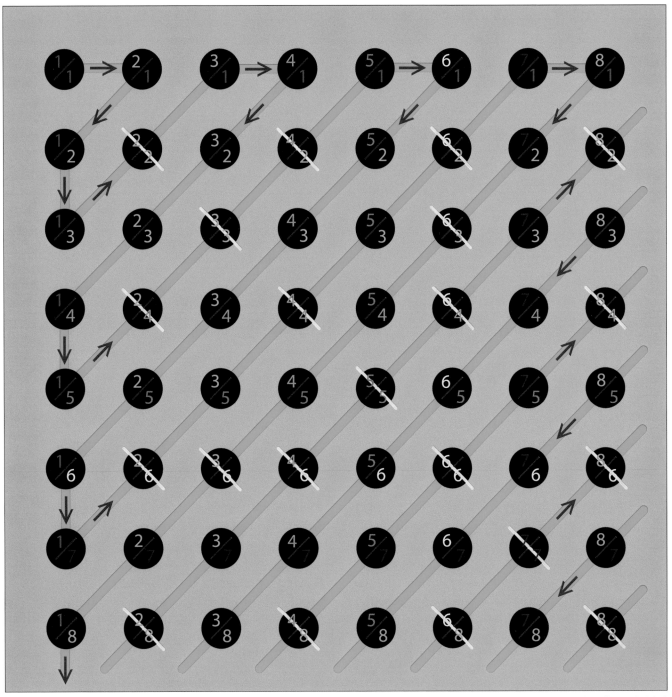

Plate 51. *The Rationals Are Countable!*

finished product. For example, the act of counting the natural numbers 1, 2, 3, . . . can go on forever, yet at each stage only a finite number of them have been counted; the complete process can never be finished. Cantor thought otherwise: the natural numbers, he said, should be regarded as a single, complete object, symbolized by enclosing them in braces: {1, 2, 3,...}. In effect, Cantor declared that infinity, far from being the elusive goal of a never-ending process, is a mathematical reality.

But this was only the harbinger of things to come. Consider the set of rational numbers. We know that between any two fractions, no matter how close, you can always squeeze a third fraction. For example, between 1/1,001 and 1/1,000 you can fit the fraction 2/2,001 and, indeed, infinitely many more fractions. The rational numbers, then, are spread densely along the number line, and it would seem only natural to assume that there are many more of them than counting numbers—infinitely many more. But "natural" is a very poor guide when it comes to infinity. In a series of groundbreaking papers published in the decade 1874–1884, Cantor established three results that seem to defy common sense:

1. There are just as many positive rational numbers as counting numbers; that is to say, we can enlist all positive rationals in a row and count them one by one, without leaving a single one out, as depicted in plate 51. Cantor called any such set a *countable*, or *denumerable*, set. The even numbers, the odd numbers, the squares, the primes, and the rational numbers are all denumerable sets: their members can be put in a one-to-one correspondence with the counting numbers. Cantor denoted their infinity with the symbol $\aleph_0$, the infinity of countable sets ($\aleph$, pronounced "aleph," is the first letter of the Hebrew alphabet).

2. The *real* numbers, on the other hand, are *uncountable*—there are infinitely many more of them than there are counting numbers or rational numbers. They cannot be put in a 1:1 correspondence with any denumerable set; their infinity is of a higher rank than that of denumerable sets. Cantor assigned this kind of infinity the Gothic letter $\mathfrak{C}$, the infinity of the *continuum*. In the hierarchy of infinities, $\mathfrak{C} > \aleph_0$.

3. There exist sets still more numerous than the reals and, therefore, of a higher degree of infinity than even $\mathfrak{C}$. Consider the *set of all subsets* of a given set—its *power set*. For example, starting with the two-element set $\{a, b\}$, we can create a new set that has all these elements as subsets: $\{\{a\}, \{b\}, \{a, b\}, \{ \}\}$ (note that we included the empty set, { }, among the subsets). This new set has $4 = 2^2$ elements. We can now repeat the process with this new set, getting a power set of $16 = 2^4$ elements. The number of subsets in this process grows very fast: the next power set will have $2^{16} = 65,536$ elements, and the one after that, $2^{65,536}$ elements, approximately 1 followed by 19,728 zeros. The process can go on forever, generating ever-larger sets of enormous, yet still finite, size.

Cantor now imagined that we can do the same with infinite sets, generating a never-ending chain of power sets, each larger than its predecessor. Each of these sets stands one rung higher on the ladder of infinities. Cantor pointed out that these sets are purely a creation of the mind; they cannot be constructed in any "real" sense; they reside in the ethereal sphere of abstract mathematics. And yet their existence is as real as that of any other mathematical object, concrete or abstract.[1]

Cantor's last years were not happy ones. Stung by the relentless opposition to his radical ideas and suffering from repeated spells of depression, he spent his final years in a mental institution, where he died in 1918. Yet his ideas slowly gained acceptance. In a way, he accomplished the vision of William Blake's famous verse in *Auguries of Innocence*:

To see the world in a grain of sand,
And heaven in a wild flower.

Hold infinity in the palm of your hand,
And eternity in an hour

NOTE:

1. For a more complete account of Cantor's theory, see Maor, *To Infinity and Beyond*, chapters 9 and 10 (Princeton, NJ: Princeton University Press, 1991).

Appendix

PROOFS OF SELECTED THEOREMS MENTIONED IN THIS BOOK

QUADRILATERALS (PAGE 6)

We refer to figure A.1. Let $\square ABCD$ denote the area of quadrilateral $ABCD$, $\triangle ABD$, the area of triangle ABD, etc. Because triangle APS has half the base length and half the height of triangle ABD, we have $\triangle APS = \frac{1}{4}\triangle ABD$. Similarly, $\triangle CQR = \frac{1}{4}\triangle CBD$. Thus,

$$\triangle APS + \triangle CQR = \frac{1}{4}\triangle ABD + \frac{1}{4}\triangle CBD = \frac{1}{4}\square ABCD.$$

By the same argument,

$$\triangle BPQ + \triangle DSR = \frac{1}{4}\triangle BAC + \frac{1}{4}\triangle DAC = \frac{1}{4}\square ABCD.$$

Therefore,

$$\square PQRS = \square ABCD - [\triangle APS + \triangle CQR + \triangle BPQ + \triangle DSR]$$

$$= \square ABCD - \frac{1}{2}\square ABCD = \frac{1}{2}\square ABCD.$$

The proof still holds when the quadrilateral has the shape of a dart (see figure A.2), except that some of the internal areas must be subtracted rather than added. We leave the proof of this case to the reader.

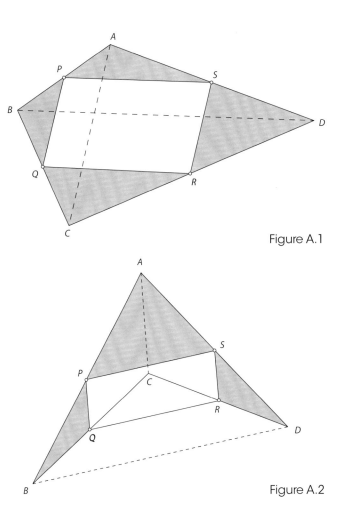

Figure A.1

Figure A.2

PYTHAGOREAN TRIPLES (PAGE 20)

Let u and v be any positive integers, with $u > v$. Set $a = u^2 - v^2$, $b = 2uv$, and $c = u^2 + v^2$. Then

$$
\begin{aligned}
a^2 + b^2 &= (u^2 - v^2)^2 + (2uv)^2 \\
&= u^4 - 2u^2v^2 + v^4 + 4u^2v^2 \\
&= u^4 + 2u^2v^2 + v^4 = (u^2 + v^2)^2 = c^2,
\end{aligned}
$$

showing that (a, b, c) is a Pythagorean triple. For example, if $u = 5$ and $v = 3$, we get $a = 5^2 - 3^2 = 25 - 9 = 16$, $b = 2 \times 5 \times 3 = 30$, and $c = 5^2 + 3^2 = 25 + 9 = 34$, giving us the triple $(16, 30, 34)$.

In this example we got a nonprimitive triple, being merely a magnification of the triple $(8, 15, 17)$ by a factor of 2. To get *primitive* triples, two restrictions must be imposed on u and v: they must be relatively prime (having no common factors other than 1), and they must be of opposite parities— one even and the other odd. For example, choosing $u = 5$ and $v = 4$ gives us the primitive triple $(9, 40, 41)$.

Since the choice of u and v is otherwise arbitrary, we see that there are infinitely many primitive Pythagorean triples. The converse is also true: every primitive Pythagorean triple corresponds to a pair of values (u, v) subject to the restrictions mentioned previously. For the details, see Maor, *The Pythagorean Theorem: A 4,000-Year History*, pp. 221–22.

A PROOF THAT $\sqrt{2}$ IS IRRATIONAL (PAGE 23)

The proof is by the indirect method: we assume that $\sqrt{2}$ is rational and then show that this leads to a contradiction.

Suppose $\sqrt{2}$ is rational, that is, $\sqrt{2} = m/n$, where m and n are two positive integers with no common factor other than 1. Rewriting this equation as $m = \sqrt{2}n$

and squaring both sides, we get $m^2 = 2n^2$, which means that m^2—and, therefore, m—is an even integer (since the square of an odd integer is always odd). So $m = 2r$ for some integer r. Putting this back into the equation $m^2 = 2n^2$, we get $4r^2 = 2n^2$, or $n^2 = 2r^2$. But this means that n^2—and, therefore, n—is also even, so $n = 2s$ for some integer s. Thus, m and n have a common factor 2, in contradiction to our assumption that the fraction m/n is in lowest terms. We conclude that $\sqrt{2}$ is not rational—QED.

EUCLID'S PROOF OF THE INFINITUDE OF THE PRIMES (PAGE 44)

We again follow an indirect proof (a proof by contradiction). Assume there is only a finite number of primes, $p_1, p_2, p_3, \ldots, p_n$ Now consider the number $N = p_1 \cdot p_2 \cdot p_3 \cdots \cdots p_n + 1$. This number is obviously greater than any of the p_i, yet it is not divisible by any of them, because any such division will leave a remainder 1. Therefore, N must either be a new prime, not included in the original list, or else it is composite, in which case at least one of its prime factors must be a new prime not included in the original list. In either case we have a contradiction, for we assumed that the set $p_1, p_2, p_3, \ldots, p_n$ included *all* the primes—QED.

To illustrate, suppose there are only three primes, 2, 3, and 5. Then $N = 2 \cdot 3 \cdot 5 + 1 = 31$, a new prime. On the other hand, if we started with 3, 5, and 7, we would get $N = 3 \cdot 5 \cdot 7 + 1 = 106 = 2 \cdot 53$, producing two new primes that were not in the original set. We can then add these primes to our list and start the process again, generating ever more primes (although not *all* of them, and not in sequential order), like a runaway nuclear chain reaction, but fortunately with less-dire consequences.

THE SUM OF A GEOMETRIC PROGRESSION (PAGE 51)

Consider the finite geometric progression a, ar, $ar^2, \ldots, ar^{n-1}$ of n terms. Let its sum be S:

$$S = a + ar + ar^2 + \cdots + ar^{n-1}.$$

Multiply this equation by the common ratio r:

$$Sr = ar + ar^2 + ar^3 + \cdots + ar^n.$$

Subtract the second equation from the first: all terms cancel out except the first and last:

$$S - Sr = a - ar^n$$

or $S(1-r) = a(1-r^n)$, from which we get

$$S = \frac{a(1-r^n)}{1-r},$$

provided, of course that $r \neq 1$. Now, if r is less than 1 in absolute value (that is, $-1 < r < 1$), the term r^n gets smaller and smaller as n increases; that is, $r^n \to 0$ as $n \to \infty$. Thus, the sum of the infinite progression is

$$S_\infty = \frac{a}{1-r}.$$

THE SUM OF THE FIRST n FIBONACCI NUMBERS (PAGE 64)

Let $F_1 = F_2 = 1$, $F_i = F_{i-2} + F_{i-1}$ for $i = 3, 4, \ldots$. Note that

$$F_1 = F_3 - F_2, \ F_2 = F_4 - F_3, \ F_3 = F_5 - F_4, \ldots, F_n = F_{n+2} - F_{n+1}.$$

Adding up these expressions, we get, on the left side, $F_1 + F_2 + F_3 + \cdots + F_n$, while on the right side all terms except $-F_2$ and F_{n+2} cancel out. Remembering that $F_2 = 1$, we thus get

$$F_1 + F_2 + F_3 + \cdots + F_n = F_{n+2} - 1.$$

A slightly different formula holds for the sum of the *squares* of the first n Fibonacci numbers:

$$F_1^2 + F_2^2 + F_3^2 + \cdots + F_n^2 = F_n F_{n+1},$$

as can be proved by mathematical induction.

Many more Fibonacci-related formulas can be found in Alfred S. Posamentier and Ingmar L. Lehmann, *The (Fabulous) Fibonacci Numbers* (Amherst, NY: Prometheus Books, 2007).

CONSTRUCTION OF A REGULAR PENTAGON (PAGE 70)

We first show that the 72-72-36-degree triangle formed by any side of the pentagon and the vertex opposite it has a side-to-base ratio equal to the golden ratio $\varphi = (1 + \sqrt{5})/2$.

Let the triangle be ABC, with the $36°$ angle at C (figure A.3). Bisect angle A and extend the bisector until it meets side BC at D. This produces two isosce-

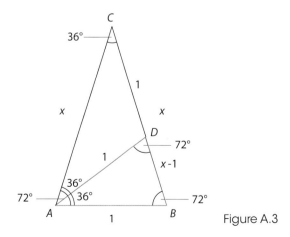

Figure A.3

les triangles—the 72-72-36-degree triangle BDA and the 36-36-108-degree triangle ACD. Of these, the former is similar to triangle ABC, so we have

$$\frac{\overline{AB}}{\overline{AC}} = \frac{\overline{BD}}{\overline{BA}}$$

(all line segments are nondirectional, so that $\overline{AB} = \overline{BA}$). Setting $\overline{AB} = 1$ and $\overline{AC} = x$, we have $\overline{AB} = \overline{AD} = \overline{CD} = 1$, $\overline{AC} = \overline{BC} = x$, and $\overline{BD} = x - 1$. We thus get

$$\frac{1}{x} = \frac{x-1}{1},$$

which leads to the quadratic equation $x^2 - x - 1 = 0$. Solving it and taking only the positive solution (because x stands for length, which cannot be negative), we get

$$x = \frac{1+\sqrt{5}}{2},$$

which is exactly the golden ratio φ.

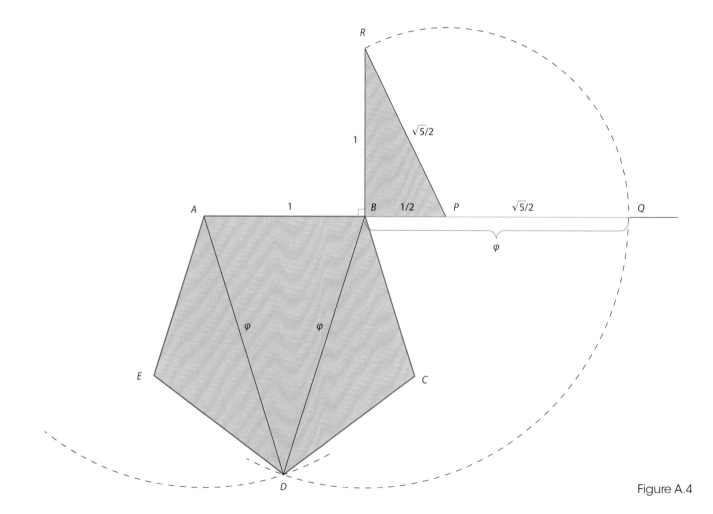

Figure A.4

To construct a pentagon with unit side, we first need to construct a segment equal in length to the golden ratio. Referring to figure A.4, draw a unit line segment $\overline{AB} = 1$, extend it to the right by segment $\overline{BP} = 1/2$, and at B erect a perpendicular $\overline{BR} = 1$. We have $\overline{PR}^2 = \overline{BP}^2 + \overline{BR}^2 = (1/2)^2 + 1^2 = 5/4$, so $\overline{PR} = \sqrt{5}/2$. Place your compass at P and swing an arc of radius $\overline{PR}$, meeting the extended segment $\overline{AB}$ at Q. We have

$$\overline{BQ} = \overline{BP} + \overline{PQ} = \frac{1}{2} + \frac{\sqrt{5}}{2} = \frac{(1+\sqrt{5})}{2} = \varphi.$$

We are now ready to construct our pentagon. Place your compass at B and swing an arc of radius $\overline{BQ}$; then do the same with your compass at A (only the first of these arcs is shown in full in figure A.4). The two arcs intersect at D. Next, from each of the points A and D swing an arc of radius $\overline{AB}$; the two arcs intersect at E (see following note). Do the same from points B and D, producing point C. Connect points A, B, C, D and E, and your pentagon is complete.

Note: for the arcs to intersect, each radius must be greater than half the segment. Indeed, $\overline{AD} = \overline{BQ} = (1+\sqrt{5})/2 \approx 1.618$, so $\overline{AE} = 1 > \overline{AD}/2 \approx 0.809$, ensuring that the arcs meet.

For further discussion and alternative ways to construct a pentagon, see Hartshorne, *Geometry: Euclid and Beyond*, chapter. 4, and Heilbron, *Geometry Civilized*, pp. 221–228.

CEVA'S THEOREM (PAGE 105)

Let the triangle be ABC (figure A.5), and let A', B', and C' be any points on the sides opposite to A, B, and C such that AA', BB', and CC' pass through one point G. Ceva thought of this point as the center of gravity of weights W_A, W_B, and W_C attached to the respective vertices. Let us find the center of gravity of any two of these weights, say W_B and W_C; call this point A'. By Archimedes's law of the lever (the "seesaw rule"), we have

$$W_B \cdot \overline{BA'} = W_C \cdot \overline{A'C},$$

from which we get

$$\frac{\overline{BA'}}{\overline{A'C}} = \frac{W_C}{W_B}.$$

In a similar manner we get for the other two pairs of points,

$$\frac{\overline{CB'}}{\overline{B'A}} = \frac{W_A}{W_C} \quad \text{and} \quad \frac{\overline{AC'}}{\overline{C'B}} = \frac{W_B}{W_A}.$$

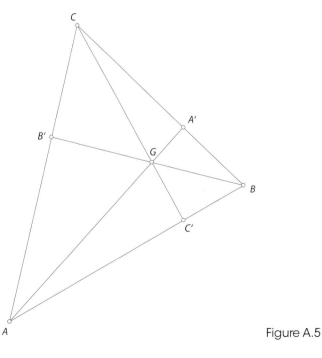

Figure A.5

Multiplying the three ratios, we get

$$\frac{\overline{BA'}}{\overline{A'C}} \cdot \frac{\overline{CB'}}{\overline{B'A}} \cdot \frac{\overline{AC'}}{\overline{C'B}} = \frac{W_C}{W_B} \cdot \frac{W_A}{W_C} \cdot \frac{W_B}{W_A} = 1.$$

Note that in forming this triple product, we went in a counterclockwise direction around the triangle, from B to A' to C, then to B', and so on. Since the center of gravity G of the entire triangle must lie on each of the lines AA', BB', and CC', it must lie on their intersection, so the three lines are concurrent. Interestingly, the actual weights disappeared from the final result—they just served as a tool and canceled out at the end.

This, of course, is a physically motivated proof with which a pure mathematician might disagree. But let us remember that Archimedes, the quintessential pure mathematician, often used subtle physical reasoning in his proofs, yet he always supplemented them with a rigorous mathematical argument. For a strictly geometric proof, see Eves, *A Survey of Geometry*, pp. 247–48.

SOME PROPERTIES OF INVERSION (PAGE 128)

Let the circle of inversion be c, with center at O and radius 1 (figure A.6). We first show that inversion transforms a straight line *not* through O into a circle *through* O. Let the line be l. Choose two points on l, the point P closest to O and any other point Q, and let their images under inversion be P' and Q', respectively. By the definition of inversion we have $\overline{OP'} = 1/\overline{OP}$ and $\overline{OQ'} = 1/\overline{OQ}$, so $\overline{OP} \cdot \overline{OP'} = \overline{OQ} \cdot \overline{OQ'} = 1$, from which we get

$$\frac{\overline{OP}}{\overline{OQ}} = \frac{\overline{OQ'}}{\overline{OP'}}.$$

This means that triangles OPQ and $OQ'P'$ are similar. Since P is the point on l closest to O, line OP is perpendicular to l, so ΔOPQ is a right triangle with its right angle at P. Therefore, triangle $OQ'P'$ is also a right triangle with its right angle at Q', and this is true regardless of the position of Q on l. Thus, by the converse of Thales's theorem (see chapter 1, note 1), as Q moves along l, its image Q' describes an arc of a circle k with diameter OP'.

The converse of this property is also true: inversion carries a circle through O onto a line not through O. This can be seen by simply reading figure A.6 "backward," from circle k to line l. This is a result of

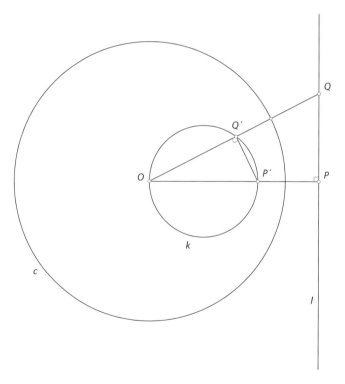

Figure A.6

the fact that inversion is completely symmetric: if P' is the image of P, then P is the image of P', as can be seen from the equivalence of the defining equations $\overline{OP'} = 1/\overline{OP}$ and $\overline{OP} = 1/\overline{OP'}$.

It can also be shown that inversion carries a circle *not* through O to another circle not through O. For the details, see Courant and Robins, *What is Mathematics,* pp. 142–144.

Bibliography

Baptist, Peter, Albrecht Beutelspacher, and Eugen Jost. *Alles ist Zahl* (*All is Number*). Cologne: Cologne University Press, 2009.

Courant, Richard, and Herbert Robbins. *What is Mathematics?* Revised ed. by Ian Stewart. New York and Oxford: Oxford University Press, 1996.

Coxeter, H.S.M. *Introduction to Geometry*. 2nd ed. New York: John Wiley, 1969.

——. *The Beauty of Geometry*. New York: Dover, 1999.

Darling, David. *The Universal Book of Mathematics: From Abracadabra to Zeno's Paradoxes*. Hoboken, N.J.: John Wiley, 2004.

Euclid: The Elements, translated with introduction and commentary by Sir Thomas Heath. 3 vols. New York: Dover, 1956.

Eves, Howard. *A Survey of Geometry*. Revised ed. Boston: Allyn and Bacon, 1972.

——. *An Introduction to the History of Mathematics*, 6th ed. Fort Worth: Saunders College Publishing, 1990.

Gombrich, E. H. *The Sense of Order: A Study in the Psychology of Decorative Art*. Ithaca, NY: Cornell University Press, 1979.

Hartshorne, Robin. *Geometry: Euclid and Beyond*. New York: Springer Verlag, 2000.

Heilbron, J. L. *Geometry Civilized: History, Culture, and Technique*. Oxford, UK: Clarendon Press, 1998.

Hidetoshi, Fukagawa, and Tony Rothman. *Sacred Mathematics: Japanese Temple Geometry*. Princeton, NJ: Princeton University Press, 2008.

Kemp, Martin. *The Science of Art: Optical Themes in Western Art from Brunelleschi to Seurat*. New Haven, CT. and London: Yale University Press, 1990.

Livio, Mario. *The Golden Ratio: The Story of Phi, the World's Most Astonishing Number*. New York: Broadway Books, 2002.

Mankiewicz, Richard. *The Story of Mathematics*. Princeton, NJ: Princeton University Press (ND).

Maor, Eli. *The Pythagorean Theorem: A 4,000-Year History*. Princeton, NJ: Princeton University Press, 2007.

——. *Trigonometric Delights*. Princeton, NJ: Princeton University Press, 2002.

——. *e: the Story of a Number*. Princeton, NJ: Princeton University Press, 2009.

Pickover, Clifford A. *The Math Book: From Pythagoras to the 57th Dimension, 250 Milestones in the History of Mathematics*. New York and London: Sterling, 2009.

Posamentier, Alfred S., and Ingmar Lehmann. *The (Fabulous) Fibonacci Numbers*. New York: Prometheus Books, 2007.

Rademacher, Hans, and Otto Toeplitz. *The Enjoyment of Mathematics*. Princeton, NJ: Princeton University Press, 1957.

Seymour, Dale, and Reuben Schadler. *Creative Constructions*. Oak Lawn, IL: Ideal School Supply Company, 1994.

Steinhaus, H. *Mathematical Snapshots*. New York: Oxford University Press, 1969.

Stevens, Peter S. *Handbook of Regular Patterns: An Introduction to Symmetry in Two Dimensions*. Cambridge, MA. and London: MIT Press, 1981.

Wells, David. *The Penguin Dictionary of Curious and Interesting Geometry*. London: Penguin Books, 1991.

Weyl, Hermann, *Symmetry*. Princeton, NJ: Princeton University Press, 1952.

WEB SITES

Cut The Knot: http://www.cut-the-knot.org/front.shtml.

Eugen Jost on the web: http://www.everything-is-number.net.

The MacTutor History of Mathematics Archive, created by John J. O'Connor and Edmund F. Robertson, The School of Mathematics and Statistics, University of St. Andrews, Scotland: http://www-history.mcs.st-andrews.ac.uk/history/.

Wolfram MathWorld: http://mathworld.wolfram.com/.

Index